Lecture Notes in Computer Science

Edited by G. Goos, J. Hartmanis and J. van Leeuwen

Advisory Board: W. Brauer   D. Gries   J. Stoer

885

Lecture Notes in Computer Science

Edited by G. Goos, J. Hartmanis and J. van Leeuwen

Advisory Board: W. Brauer D. Gries J. Stoer

Remco C. Veltkamp

Closed Object Boundaries from Scattered Points

Springer-Verlag
Berlin Heidelberg New York
London Paris Tokyo
Hong Kong Barcelona
Budapest

Series Editors

Gerhard Goos
Universität Karlsruhe
Vincenz-Priessnitz-Straße 3, D-76128 Karlsruhe, Germany

Juris Hartmanis
Department of Computer Science, Cornell University
4130 Upson Hall, Ithaka, NY 14853, USA

Jan van Leeuwen
Department of Computer Science, Utrecht University
Padualaan 14, 3584 CH Utrecht, The Netherlands

Author

Remco C. Veltkamp
Department of Interactive Systems, CWI
Kruislaan 413, 1098 SJ Amsterdam, The Netherlands

CR Subject Classification (1991): I.3.5, G.2.2, I.4.8, I.5.4

ISBN 3-540-58808-6 Springer-Verlag Berlin Heidelberg New York

CIP data applied for

© Springer-Verlag Berlin Heidelberg 1994
Printed in Germany

Typesetting: Camera-ready by author
SPIN: 10479073   45/3140-543210 - Printed on acid-free paper

Preface

This Ph.D. dissertation presents the result of research carried out between 1985 and 1992, first at Leiden University as a scientific assistant researcher, and later at CWI (Centre for Mathematics and Computer Science), Amsterdam, as a researcher on the NFI IIICAD project, funded by NWO (Dutch Organization for Scientific Research) under Grant NF-51/62-514.

The research goal was the development of new methods and techniques for the construction of closed object boundaries from scattered points in both 2D and 3D. These points are either synthetic or measured from the boundary of an existing object.

New results are presented in Chapters 3, 5, 7, and 9. Chapter 3 introduces 'the γ-neighborhood graph', which provides a geometrical structure on the scattered points. Chapter 5 presents a method to construct a piecewise linear boundary through all given scattered points which is based on the γ-neighborhood graph. Chapter 7 introduces 'the flintstones scheme', a hierarchical approximation and localization scheme. Chapter 9 presents methods to construct a smooth piecewise cubic boundary from a piecewise linear one (e.g. resulted from methods of Chapter 5 or 7).

The material presented in this dissertation has partly appeared before in other publications. The correspondence between the chapters and publications is as follows: Chapters 2 and 3: [Veltkamp, 92c], Chapters 4 and 5: [Veltkamp, 89a] and [Veltkamp, 91], Chapters 6 and 7: [Veltkamp, 90] and [Veltkamp, 92b], Chapter 8: [Veltkamp, 92d], Chapter 9: [Veltkamp, 92a].

The text of this thesis could never have matured without the suggestions and constructive criticism of many colleagues, above all my Ph.D. supervisors Jan van den Bos and Mark Overmars. Arie de Bruin, Nies Huijsmans, Pia Pfluger,

Roel Stroeker, and Cees Traas made valuable comments on this manuscript. In addition, Nies Huijsmans played a crucial role in stimulating me at hard times. Several people contributed to the development of first versions of chunks of software: my colleagues Peter van Oosterom and Rene Pluis, and the students Timo Koornstra, Roel van der Land, and Jos van Hillegersberg. Paul ten Hagen made it possible to finish this research as part of the NFI IIICAD project. Finally, Desiree Capel helped with the production of the manuscript through mental support and professional editorial remarks.

Contents

1

Introduction

The work described in this thesis deals with the computational processing of various forms of geometric information, and could therefore be considered research in Computational Geometry. This is correct if the term Computational Geometry is used in a broad sense, but several authors use this term for more specific subjects. It has been used to label shape recognition on parallel machines [Minsky and Papert, 69], design and manufacturing of 2D and 3D shapes [Forrest, 71], the computational aspects of integral and stochastic geometry [Bernroider, 78], curve and surface modeling [Su and Liu, 89], and combinatorial and algorithmic issues in discrete geometry [Shamos, 78].

To date, the latter meaning of Computational Geometry is most widely used, comprising the area that deals with problems concerning points, lines, polygons, planes, polyhedra, and so on. Applications of discrete Computational Geometry can be found in computer graphics, computer vision, robot motion planning, and VLSI design. Of the following chapters, Chapter 2 to 7 fall in the category of discrete Computational Geometry.

Curve and surface modeling, i.e. continuous Computational Geometry, is commonly referred to as (curves and surfaces in) Computer Aided Geometric Design (CAGD). Examples of topics in this field are the interpolation and approximation with B-splines and Bézier curves and surfaces, and Coons patches and similar schemes. Applications of CAGD are widely used in the automobile, aircraft, and ship-building industry. Chapters 8 and 9 fall in the category of continuous Computational Geometry.

The work of this thesis is more specifically concerned with the computational aspects of geometry with respect to form or shape information, that is, morphology. Indeed, morphology is closely related to geometry, and thus to

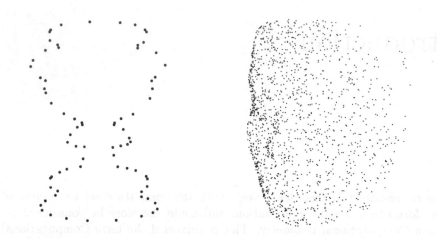

Figure 1.1. 2D boundary points of a chalice, and 3D boundary points of a mask.

Computational Geometry. A computational geometric approach to the analysis of form is called Computational Morphology [Toussaint, 88a]. In particular, this thesis is about the construction of closed object boundaries from scattered points.

1.1 Boundary construction

In many applications in geometric modeling, computer graphics, object recognition, distance map image processing, and computer vision, input data is available in the form of a set of 2D or 3D coordinates that are points on the boundary of an object. Such points can be synthetic or measured from the boundary of an existing object. See Figure 1.1 for an example of a set of 2D points from the boundary of Uccello's chalice, which serves as the cover picture of the journal 'Computer Aided Geometric Design' [Thoenes, 84], and a collection of 3D points from the boundary of a mask, measured by a laser range system [Rioux and Cournoyer, 88]. A collection of points, however, is an ambiguous representation of an object, and can therefore not be used directly in many applications. It is often essential to have a representation of the whole boundary available, which is unambiguously defining a valid object. The boundary constructed from a set of points can for example be used for the initial design of an artifact, for numerical analysis, or for graphical display.

The way in which the boundary points are acquired may give useful information in order to construct the whole boundary, but can also make the construction method very dependent on the specific data source. If it is not known how the data is obtained or if a single construction method is to be used for data from various types of sources, no structural relation between the input points may be

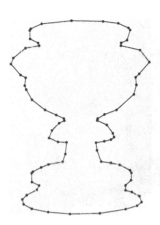

Figure 1.2. Polygonal chalice boundary, and polyhedral mask boundary.

assumed, except that they all lie on the boundary of an object. The order of the points in the input then provides no information on their topological relation to each other. In particular, they do not lie on a regular grid, but are scattered points. This thesis is about the development of new techniques to construct and manipulate closed boundaries of 2D and 3D objects from scattered points.

The simplest boundary through a set of points is one that consists of linear segments: line segments for a 2D polygonal boundary, and triangles for a 3D polyhedral boundary (in 3D, a triangle is the unique polygon that is always flat). Figure 1.2 shows a polygonal and polyhedral boundary of the points from Figure 1.1. In both the polygonal and the polyhedral boundary, points are connected by edges. Trying all possible boundaries through a given set of points by considering all edges between points is not feasible because of the combinatorial explosion of the number of possible solutions. For example in 2D, a boundary through N_v points must consist of N_v line segments, and there are $\binom{N_v}{2}$ possible edges. Trying all sets of N_v edges out of $\binom{N_v}{2}$ results in as many as

$$\Theta\left(\binom{\binom{N_v}{2}}{N_v}\right)$$

combinations, which is too much to be of practical use. One possible solution to this problem is to first describe some structure of the set of points by a geometric graph, and then derive a boundary from this structure using the inherent geometric information. This approach is taken in this thesis.

In many real applications, a boundary constructed from a set of points consists of thousands of faces. For example, the constructed boundary of the mask in Figure 1.2 consists of about three thousand triangles. However, an approximation of the object is often sufficient. In animation for example, the motion

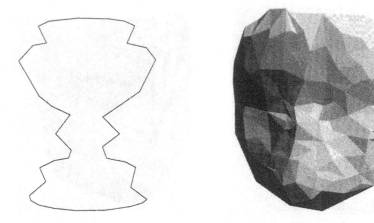

Figure 1.3. Polygonal and polyhedral approximations.

blur prohibits the perception of much detail, so that an approximated object is sufficient and is also faster to display. A polygonal approximation of the chalice and a polyhedral approximation of the mask are shown in Figure 1.3.

Localization provides bounding area or volume information. Such information is useful for efficient operations such as collision detection for robot motion planning. Because boundaries constructed from experimental data often consist of many segments, and because a hierarchy of approximations together with localization information is very efficient for many applications, this subject is also treated in this thesis.

Our goal is to devise a scheme whose definition is readily generalized from 2D to 3D, and is very efficient in use. This is not as easily obtained as it may seem, for many existing methods do not meet these demands. For example, a simple bounding area for a piece of a polygon is a rectangle. The generalization to 3D suggests the use of a block, but the intersection test for two blocks may require thirty-six intersection tests between the sides of the blocks. On the other hand, the test for intersection of two circles or spheres only requires the calculation of the distance between the two centers: if the distance is smaller than the sum of the two radii, there is an intersection. Indeed, circles and spheres are used as bounding areas and volumes since the early days of geometric modeling and computer graphics. The application of circles and spheres to the approximation and localization of polygons and triangular polyhedra in a hierarchical way is elaborated in this thesis.

Polygonal boundaries are C^1-discontinuous at the vertices, exhibiting abruptly changing directions of the tangent line. Given an ordered set of vertices, i.e. a polygon, a smoother boundary curve is often desired, consisting of curved line segments that interpolate the vertices of the straight line segments and are smoothly connected at the vertices. Analogously, the polyhedral boundaries

Figure 1.4. Smooth chalice curve and mask surface.

are C^1-discontinuous at the edges, where the tangent planes instantly change orientation. A smoother surface, consisting of curved triangles that interpolate the flat triangles' vertices and are smoothly connected along the edges, is often desired. For example, esthetic demands apply to car body design, physical requirements deduced from aerodynamic or hydrodynamic laws apply to aircraft and ship hull design, and for boundary reconstruction the smoothness demands are determined by the smoothness of the original boundary. Figure 1.4 shows a tangent line continuous chalice boundary curve that interpolates the vertices of the polygonal boundary, and a tangent plane continuous mask boundary surface that interpolates the triangle vertices of the polyhedral boundary.

Smooth boundaries are most easily constructed by piecewise polynomials. Three polynomial schemes are most widely used in Computer Aided Geometric Design, the Coons, B-spline and Bézier schemes, especially for rectangular surface patches. Triangular interpolants are dominant in 3D scattered data interpolation [Barnhill, 85], and the Bézier formulation for curves naturally generalizes to a triangular form. (Bézier curves and surfaces were independently developed by de Casteljau at the Citroën and by Bézier at the Renault automobile company, but de Casteljau's development was never published, so that this curve and surface scheme was named after Bézier.) The Bézier formulation is a convenient method to describe other polynomial schemes as well as to develop new schemes.

Because the Bézier scheme has useful geometrical interpretations, and results in a piecewise, triangle by triangle, surface representation, the interpolation methods developed in this thesis are in Bézier form. The problem treated in this thesis is the development of a tangent line/plane continuous interpolation scheme that is local, i.e. only depends on nearby vertices, while keeping the polynomial degree as low as three. This is easily done for curves, but is more involved for surfaces. Indeed, the polynomial degree of tangent plane continuous triangular surfaces is usually four [Piper, 87] or five [Pfluger and Neamtu, 91].

1.2   Outline of the thesis

Four topics are treated in this thesis: geometric graphs, piecewise linear boundary construction, hierarchical approximation and localization, and smooth boundary construction. Chapter 2 introduces some concepts that are used throughout this thesis, and presents an introduction to geometric graphs. In particular an overview of geometric graphs that describe some internal or external structure of a set of points is presented. Chapter 3 introduces a new geometric graph: the γ-neighborhood graph. The way in which the internal structure of a set of points is described by the γ-neighborhood graph is used to derive the topological relation between the points, assuming that they are on the boundary of an object.

Chapter 4 states the precise boundary construction problem, motivates the use of a suitable geometric graph, and gives an overview of existing methods to solve the problem. In Chapter 5, the γ-neighborhood graph is used for boundary construction, which leads to advantages over other methods.

Chapter 6 gives a more detailed explanation of the merits of (hierarchical) approximation and localization, and an overview of existing schemes. Since the various methods use approximation error criteria that are closely related to the methods, several error criteria are introduced as well. Chapter 7 introduces a new method, the flintstones scheme, a scheme for both 2D and 3D, based on a bounding area/volume defined by circles/spheres, which makes this scheme efficient for use in hierarchical operations.

Chapter 8 introduces the relevant concepts for smooth boundaries, like Bézier curves and surfaces, and geometric (G^n-) continuity. An obvious way to generate a smooth boundary curve in 2D is presented in Chapter 9. This chapter further presents an analysis of the sufficient and necessary polynomial degree for several smooth interpolation problems in 3D, and introduces three new schemes for the construction of a smooth piecewise cubic Bézier surface.

1.3   Conventions

Throughout this thesis N_v denotes the number of vertices, N_e the number of edges, and N_t the number of triangles that are considered. All distances are Euclidean, or L_2-, distances.

The notation '2D' means 'two-dimensional', or 'two-dimensional space', and likewise for '3D' and 'kD'. The word 'boundary' is used for a 2D boundary curve as well as for a 3D boundary surface, and other expressions like 'boundary segment' and '(boundary) simplex' are used in the same way. Many statements about such elements hold in both the 2D and the 3D situation, so that the dimensionality is often omitted.

To give orders of time and storage complexities, we employ the commonly used notation of [Knuth, 76]:

DEFINITION 1.1 (ORDERS OF COMPLEXITY)

$\Theta(f(N))$ *denotes the set of all $g(N)$ such that there exist positive constants c_1, c_2, and N_0 with $c_1 f(N) \leq g(N) \leq c_2 f(N)$ for all $N \geq N_0$,*

$\mathcal{O}(f(N))$ *denotes the set of all $g(N)$ such that there exist positive constants c and N_0 with $|g(N)| \leq c f(N)$ for all $N \geq N_0$,*

$\Omega(f(N))$ *denotes the set of all $g(N)$ such that there exist positive constants c and N_0 with $g(N) \geq c f(N)$ for all $N \geq N_0$,*

where N is the size of the input of the problem.

$\Theta(f(N))$ can be read as 'order exactly $f(N)$', $\mathcal{O}(f(N))$ as 'order at most $f(N)$', which gives an upper bound, and $\Omega(f(N))$ as 'order at least $f(N)$', which gives a lower bound, all three 'for large N'.

We can say for example that the worst case time complexity of a given algorithm is $\Theta(f(N))$. Note that this is a stronger statement than saying that for an arbitrary case the time complexity is $\mathcal{O}(f(N))$, which would not imply that the order exactly $f(N)$ is actually reached.

The definitions above refer to 'the set of all $g(n)$ ...' rather than to 'an arbitrary function $g(n)$ ...'. The phenomenon of one-way equalities arises here: $\mathcal{O}(f(n)) = \mathcal{O}(g(n))$ actually means $\mathcal{O}(f(n)) \subseteq \mathcal{O}(g(n))$, i.e. set inclusion. The use of a one-way equality instead of inclusion for the Ω-, $\mathcal{O}$-, and Θ-notation has become common practice .

2

Geometric graphs

This chapter presents a number of concepts that will be used throughout this thesis. After a few elements of general graph theory have been introduced, an overview of geometric graphs and their interrelationships will be given.

2.1 Introduction

A graph is a structure that allows the representation of the existence of a relation between elements. Formally, a graph is defined as follows:

DEFINITION 2.1 *A graph G is a pair (V, E) where V is a non-empty finite set of N_v distinct elements $v_0, \ldots, v_{N_v-1}$, and E is a set of unordered sets $\{v_i, v_j\}$, $0 \le i, j \le N_v - 1$, $i \ne j$.*

The elements of V are called vertices, the elements of E edges. We will simply denote an edge $\{v_i, v_j\}$ with '$v_i v_j$'. An edge $v_i v_j$ represents the existence of a relation between v_i and v_j. $v_i v_j$ and $v_j v_i$ mean the same edge; edges are undirected. If $v_i v_j \in E$, then v_i and v_j are *adjacent*, or *neighbors*, and incident to $v_i v_j$. Two edges are adjacent if they have a common vertex.

In a *spanning* graph all vertices are incident to an edge. A graph $G = (V, E)$ is called *empty* if $E = \phi$, and *complete* if E is the set of all possible edges. A graph $G' = (V', E')$ is called a *subgraph* of G if $V' \subseteq V$ and $E' \subseteq E$. The union of graphs (V, E) and (W, F) is $(V, E) \cup (W, F) = (V \cup W, E \cup F)$. A *path* is a sequence of vertices $v_{i_0} \ldots v_{i_j}$ such that each pair of consecutive vertices is an edge in the graph. We say that such a path is between v_{i_0} and v_{i_j}. A closed path is a sequence of vertices $v_{i_0} \ldots v_{i_j}$ such that $v_{i_0} \ldots v_{i_j}$ is a path and $v_{i_0} v_{i_j}$ is an edge in the graph. A closed path is a *cycle* if all its vertices are distinct. A *Hamilton* cycle is a cycle containing all vertices of the graph. A graph is *connected* if there is a path between every pair of distinct vertices. A graph is *n-connected* if there are n different paths between any two distinct vertices, or equivalently, if the removal of any $n-1$ vertices leaves the graph connected. A graph without cycles is a *forest*. A *tree* is a connected graph without cycles. One vertex of the tree can be denoted as the *root*.

A graph is called a *geometric* graph if the vertices represent points in a Euclidean space, and the edges represent some geometric relation between the vertices. The length of an edge of a geometric graph is the Euclidean distance between its two vertices. The length of a geometric graph is the sum of the lengths of all edges, and the length of a path is the sum of the lengths of the edges in that path.

A hyper-graph is a generalization of a graph ($\mathcal{P}(V)$ is the power set of V, the set of all subsets of V):

DEFINITION 2.2 *A hypergraph G is a pair (V, E) where V is a non-empty finite set of distinct elements, and E is a subset of $\mathcal{P}(V) \backslash V$ not containing ϕ.*

Let $v_0, \ldots, v_{N_v}$ be vertices in an Euclidean space. A polyline is a finite ordered sequence of line segments $v_{i_0} v_{i_1}, v_{i_1} v_{i_2}, \ldots, v_{i_{n-1}} v_{i_n}$, such that $v_{i_j} = v_{i_k}$ if and only if $j = k$. A polyline of consecutive line segments $v_{i_0} v_{i_1}, \ldots, v_{i_{n-1}} v_{i_n}$, is denoted by $v_{i_0} \ldots v_{i_n}$. Every line segment end point is shared by exactly one or two line segments.

A polygon $v_{i_0} \ldots v_{i_n}$ is a polyline $v_{i_0} \ldots v_{i_n}$ that is closed by segment $v_{i_n} v_{i_0}$. Every line segment end point is shared by exactly two line segments. A polygon is *simple* if its line segments share no points other than end points. For compatibility with the generalization to 3D, the edges of a polygon are called *faces*.

Polylines and polygons can be uniquely represented by a graph (V, E) where V is the set of vertices of that polyline or polygon and E the set of edges $v_i v_j$ that are the line segments. A polyline of N_v vertices has $N_v - 1$ edges; a polygon has N_v edges. Since every vertex is shared by one or two edges, there is a unique ordering of edges in a polyline or polygon $v_{i_0} \ldots v_{i_n}$. A triangle is the unique polygon of three non-collinear vertices, that is, there is only one set of three edges joining the three vertices.

A closed polyhedron in 3D is a finite set of plane polygons such that every line segment of a polygon is shared by exactly one other polygon, and no subset of polygons has the same property. Consequently, if line segments of a closed polyhedron have more than an end point in common, they must coincide and cannot only partially overlap. An open polyhedron is a connected subset of polygons of a closed polyhedron. A polyhedron is simple if there is no pair of non-adjacent polygons sharing a point. The polygons of a polyhedron are called faces.

In the restricted case that the polygon has no through-passages, i.e. it is topologically equivalent to a sphere, Euler's formula applies: $N_v - N_e + N_t = 2$. Since for a closed triangulation $3N_t = 2N_e$, it follows that for a polyhedron of triangles without through-passages $N_t = 2N_v - 4$ holds. A polyhedron of triangles can be uniquely represented by a hyper-graph (V, T) where V is the set of vertices and T the set of triangles $v_i v_j v_k$ of the polyhedron. A tetrahedron is the unique polyhedron of four non-coplanar vertices, that is, there is only one set of four triangles joining the four vertices.

A simplex or k-simplex is the unique kD structure of $k+1$ vertices not lying in a $(k-1)$D hyper-plane that joins its vertices by $k+1$ $(k-1)$-simplices; a 1-simplex of two vertices is a line segment between these vertices. So, a 2-simplex is a triangle and a 3-simplex is a tetrahedron.

A graph is depicted by representing a vertex by a dot, and an edge $v_i v_j$ by a line segment between the dots corresponding to v_i and v_j. Informally, a graph is *planar* if it can be drawn in the plane without crossing edges.

There is an obvious mapping from a hyper-graph (V, E) to a graph (V, E') associating all the $\binom{j}{2}$ edges of pairs of vertices in e with each element $e = v_{i_1} \ldots v_{i_j}$ of E. A hyper-graph can be displayed by depicting the resulting graph.

In the following I will often omit the word 'hyper' in terms like hyper-sphere, hyper-plane, and hyper-graph, when it is clear from the context that these generalized terms are appropriate.

More about graph theory in general can be found in [Bollobás, 79], and graphs in computational geometry in [Mehlhorn, 84], [Edelsbrunner, 87], and [Preparata and Shamos, 85].

2.2   Overview of geometric graphs

The rest of this chapter is concerned with geometric graphs. Some of the graphs mentioned in this section are neighborhood graphs (or hyper-graphs). A neighborhood associated with $v_1, \ldots, v_n$ is an open part of the embedding space; its definition depends on the particular graph but is only dependent of $v_1, \ldots, v_n$. A neighborhood graph joins vertices if the associated neighborhood is empty, that is, if no other vertices lie in the neighborhood. However, in the special case that a kD neighborhood associated with $v_1, \ldots, v_k$ is defined by a half-space with its bounding hyper-plane through $v_1 \ldots v_k$, the closed part of the hyper-plane bounded by the $(k-1)$-simplex $v_1 \ldots v_k$ is part of the neighborhood. So, a neighborhood that is a 3D half-space with its boundary through $v_1 v_2 v_3$ is empty if no other vertices lie in the open half-space, nor inside nor on triangle $v_1 v_2 v_3$, but vertices may lie on the boundary plane outside the triangle.

Some of the neighborhoods are defined in terms of discs or balls. A disc is the closed point set bounded by a circle. A disc is said to touch a vertex if the bounding circle passes through the vertex. The same applies to the kD analogues of disc and circle: ball and sphere. Occasionally the radius of a ball is given by a parameterized expression. In such expressions, $x/0 = \infty$ for $x \in \mathbb{R}\backslash\{0\}$. A kD ball of infinite radius touching $v_1, \ldots, v_k$ is considered a half-space with its boundary through $v_1, \ldots, v_k$.

2.2.1 Closest Pairs

DEFINITION 2.3 (CLOSEST PAIRS (CP)) *Let V be a set of vertices in kD. The Closest Pairs of V is the graph (V, E) with E the set of edges $v_i v_j$ such that $d(v_i, v_j) \leq d(v_k, v_\ell)$ for all v_k, v_ℓ, $v_k \neq v_\ell$.*

Note that there can be more than one edge in the graph, that is, more than one closest pair. The Closest Pairs graph is generally disconnected. The Closest Pairs of a set of vertices in kD can be found in $\Theta(N_v \log N_v)$ time, provided that the maximum number of vertices joined to each vertex is independent of N_v [Bentley and Shamos, 76]. This time complexity is optimal.

2.2.2 Nearest Neighbors Graph

DEFINITION 2.4 (NEAREST NEIGHBORS GRAPH (NNG)) *Let V be a set of vertices in kD. The Nearest Neighbors Graph of V is the graph (V, E) with E the set of edges that joins each vertex v_i with one v_j satisfying $d(v_i, v_j) \leq d(v_i, v_k)$ for all $v_k \neq v_i$.*

Note that the Nearest Neighbors Graph is not unique if there is more than one v_j such that $d(v_i, v_j) \leq d(v_i, v_k)$ for all $v_k \neq v_j$. The Nearest Neighbors Graph is generally disconnected. Since all the pairs of vertices that are each other's nearest neighbor contain the pairs with the smallest distance of all, CP $\subseteq$ NNG. The Nearest Neighbors Graph in kD can be constructed in $\Theta(N_v (\log N_v)^{k-1})$ time [Bentley and Shamos, 76], provided that the maximum number of vertices joined to each vertex is independent of N_v.

2.2.3 Euclidean Minimum Spanning Tree

DEFINITION 2.5 (EUCLIDEAN MINIMUM SPANNING TREE (EMST)) *Let V be a set of vertices in kD. A Euclidean Minimum Spanning Tree of V is a spanning tree of minimum length.*

The Euclidean Minimum Spanning Tree need not be unique. In a Euclidean Minimum Spanning Tree, each vertex must be joined to its nearest vertex, and thus NNG $\subseteq$ EMST, provided that the Nearest Neighbors Graph on the vertex set is unique. The Nearest Neighbors Graph actually is a minimum spanning forest, so in the special case that the Nearest Neighbors Graph is connected, it coincides with the Euclidean Minimum Spanning Tree. In 2D the Euclidean Minimum Spanning Tree can be found in $\Theta(N_v \log N_v)$ time [Shamos and Hoey, 75], which is optimal, and in higher dimensions in $\mathcal{O}(N_v^2)$ time [Prim, 57].

2.2.4 Infinite Strip Graph

The Infinite Strip Graph (∞-SG) joins two vertices if and only if the associated infinite strip is empty [Devroye, 88]

DEFINITION 2.6 (INFINITE STRIP) *Let v_1 and v_2 be two vertices in kD. The infinite strip is the open space bounded by two parallel $(k-1)D$ planes through v_1 and v_2 perpendicular to v_1v_2.*

The infinite strip can be considered a neighborhood, although 'neighborhood' suggest locality, while it is infinite in this case. However, the *definition* is local in the sense that it only depends on the two vertices.

If no two infinite strips for different pairs of vertices coincide, the Euclidean Minimum Spanning Tree must join a pair of vertices whose infinite strip is empty. So in non-degenerate cases ∞-SG $\subseteq$ EMST. Therefore, one can examine each of the $N_v - 1$ edges in the Euclidean Minimum Spanning Tree, and check if any vertex lies in the corresponding infinite strip. So the Infinite Strip Graph can be constructed in $\mathcal{O}(N_v^2)$.

2.2.5   Sphere of Influence Graph

The Sphere of Influence Graph was introduced by [Toussaint, 88b] for vertices in 2D. However, the definition is generalized to higher dimensions in a straight-forward manner:

DEFINITION 2.7 (SPHERE OF INFLUENCE GRAPH (SIG)) *Let V be a set of vertices in kD. For each vertex v, let r_v be the distance to its closest vertex. The Sphere of Influence Graph joins two vertices v_1 and v_2, if and only if the sphere centered at v_1 with radius r_{v_1} and the sphere centered at v_2 with radius r_{v_2} intersect in more than one point.*

The Sphere of Influence Graph may be disconnected. Clearly, each vertex is joined with its nearest neighbor, so that NNG $\subseteq$ SIG.

The 2D Sphere of Influence Graph can be constructed in $\Theta(N_v \log N_v)$ time [Toussaint, 88b], which is optimal. The higher dimensional Sphere of Influence Graph can be constructed after computing the Nearest Neighbors Graph (in order to determine the r_v of every vertex v), by examining all the $N_v(N_v - 1)/2$ pairs of vertices in constant time. So, the higher dimensional Sphere of Influence Graph can be constructed in $\mathcal{O}(N_v^2)$ time.

2.2.6   Relative Neighborhood Graph

The Relative Neighborhood Graph (RNG) joins two vertices if and only if their relative neighborhood is empty.

DEFINITION 2.8 (RELATIVE NEIGHBORHOOD) *Let v_1, v_2 be two vertices in kD. The associated relative neighborhood is the interior of the intersection of the two kD balls centered at v_1 and v_2 with radius $d(v_1, v_2)$.*

Two vertices v_1 and v_2 with an empty relative neighborhood are said to be rela-tively close, i.e. if $d(v_1, v_2) \leq \max\{d(v_1, v_i), d(v_2, v_i)\}$, for all $v_i \neq v_1, v_2$. In the original definition by [Lankford, 69], the '$\leq$' is replaced by a '$<$', but the former definition has become common in computational geometry [Toussaint, 80], and corresponds to our notion of empty neighborhood. It is shown by [Toussaint, 80] that EMST $\subseteq$ RNG.

The Relative Neighborhood Graph can be constructed in $\Theta(N_v \log N_v)$ time in 2D [Supowit, 83], and in $\mathcal{O}(N_v^3)$ time in higher dimensions [Toussaint, 80].

2.2.7 Gabriel Graph

The Gabriel Graph (GG), named after its originator [Gabriel and Sokal, 69], joins two vertices if and only if their Gabriel neighborhood is empty. It has originally been defined for 2D, but the definition is generalized to higher dimensions in a straightforward way:

DEFINITION 2.9 (GABRIEL NEIGHBORHOOD) *Let v_1, v_2 be two vertices in kD. The Gabriel neighborhood associated with v_1 and v_2 is the interior of the smallest kD ball touching v_1 and v_2.*

The Gabriel neighborhood sphere has radius $d(v_1, v_2)/2$. Because the Gabriel neighborhood is contained in the relative neighborhood, it is empty when the latter is empty, and therefore RNG $\subseteq$ GG. The Gabriel Graph has originally been used for analysis of geographic variation of data (e.g. the cubic root of the body weight of female red-winged blackbirds in North-America [Gabriel and Sokal, 69]).

The Gabriel Graph in 2D can be constructed in $\Theta(N_v \log N_v)$ time, which is optimal, see [Matula and Sokal, 80]. The higher dimensional Gabriel Graph can be constructed by brute force in $\mathcal{O}(N_v^3)$ time.

2.2.8 Convex Hull

I will define the Convex Hull in terms of a hyper-graph, but there are many other ways.

DEFINITION 2.10 (CONVEX HULL (CH)) *Let V be a set of vertices in kD. The Convex Hull of V is the hyper-graph (V, F) where F is the set of $(k-1)$-simplices $v_{i_0} \ldots v_{i_k}$ such that no other vertices lie in the open half-space at one side of the hyper-plane through $v_{i_0} \ldots v_{i_k}$, nor inside nor on simplex $v_{i_0} \ldots v_{i_k}$.*

Informally speaking, Convex Hull is the tightest hull enclosing V, which is convex indeed. However, in the above definition the faces are additionally required to be simplices. For example in 3D, the faces must be triangles, even if two adjacent triangles are coplanar and could as well be replaced by a quadrilateral. So, the Convex Hull is a polyhedron with triangular faces. Notice that the half-space in the definition can be considered a neighborhood.

The 2D and 3D Convex Hull can be constructed in $\Theta(N_v \log N_v)$ time, which is optimal [Preparata and Hong, 77], and the kD Convex Hull, $k > 3$, in $\mathcal{O}(N_v^{\lfloor k/2 \rfloor})$ time [Chazelle, 91], which is also optimal.

2.2.9 Delaunay Triangulation

[Voronoï, 08] defines a partitioning of space into simplices L_i, whose vertices are a given set of vertices V in kD. This so-called L-subdivision is the combinatorial geometrical dual (see below) of what is now commonly known as the Voronoi Diagram, or sometimes closest point Voronoi Diagram. The Voronoi Diagram consists of cells $C_i = \{x \in \mathbb{R}^k | d(x, v_i) \leq d(x, v_j), \text{ for all } j \neq i\}$, i.e. the locus of all points in space closer (or equally close) to v_i than to any other vertex. Later, the furthest-point Voronoi Diagram has been defined as the collection of cells $C_i' = \{x \in \mathbb{R}^k | d(x, v_i) \geq d(x, v_j), \text{ for all } j \neq i\}$, the locus of points in space further (or equally far) from v_i than from any other vertex.

A definition of the L-subdivision given by [Delaunay, 28] and [Delaunay, 34] defines a simplex to be part of the L-subdivision if the ball touching its vertices is empty. The L-subdivision is now commonly called Delaunay Triangulation (DT), or sometimes closest point Delaunay Triangulation, and is the dual of the closest point Voronoi Diagram. The furthest-point Delaunay Triangulation is the dual of the furthest-point Voronoi Diagram. The Delaunay Triangulation is the dual of the associated Voronoi Diagram in the sense that each j-simplex in the Delaunay Triangulation corresponds to a $(k - j)$-simplex in the Voronoi Diagram. In particular, the vertices in the Voronoi Diagram correspond to k-simplices in the Delaunay Triangulation.

In 3D we can call the Delaunay Triangulation a Delaunay *tetrahedralization*, but in general kD this subdivision into simplices is still called a triangulation.

In the degenerate case that more than $k + 1$ vertices lie on an empty sphere, joining all these vertices with each other would generate overlapping simplices. Instead of doing that, the Delaunay Triangulation arbitrarily joins these vertices so as to generate non-overlapping simplices that fill the space enclosed by the Convex Hull of these vertices. A degenerate Delaunay Triangulation is therefore not unique.

Apart from degenerate configurations, $k + 1$ vertices form a simplex if the ball touching the vertices is empty. Naturally, this ball touches the vertices of each of the $k + 1$ $(k - 1)$-simplices that constitute the k-simplex. Those $(k - 1)$-simplices that are part of two adjacent k-simplices have two such empty balls, and the balls are the largest empty balls touching the vertices of the $(k - 1)$-simplex. But also the $(k - 1)$-simplices that are part of only one k-simplex have two empty balls, one of which has an infinite radius. These $(k - 1)$-simplices are part of the Convex Hull, so CH $\subseteq$ DT. To get an intuitive picture, let $k = 3$ and read tetrahedron for k-simplex, and triangle for $(k - 1)$-simplex.

By the observation above, the Delaunay Triangulation can be defined as a hyper-graph in the following way:

DEFINITION 2.11 (DELAUNAY TRIANGULATION) *Let V be a set of vertices in kD. The Delaunay Triangulation is the hyper-graph (V, S) with S the set of $(k - 1)$-simplices that have two empty balls touching their vertices, such that S forms non-overlapping k-simplices.*

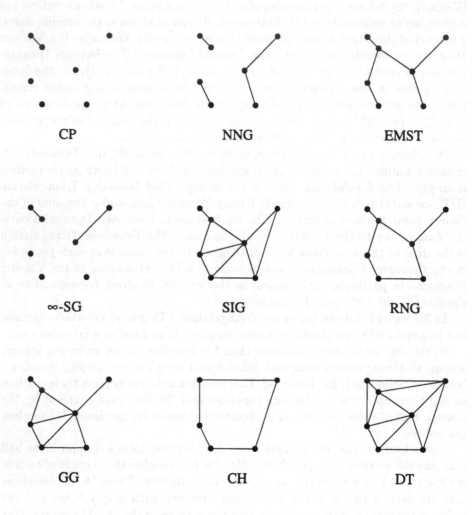

Figure 2.1. Some geometric graphs on the same set of vertices.

The interior of the two empty balls is sometimes called the Delaunay neighborhood, but note that this is only defined for $(k-1)$-simplices in the graph and not for k arbitrary vertices.

Clearly an empty ball touches the end points of each edge of a $(k-1)$-simplex in the Delaunay Triangulation. Conversely, if some ball touching two vertices v_1, v_2 is empty, there is also a largest empty ball touching these vertices, which touches $k-3$ other vertices, so that $v_1 v_2$ is an edge of a $(k-1)$-simplex that is part of the Delaunay Triangulation. So if the Gabriel neighborhood of v_1, v_2 is empty, $v_1 v_2$ is in the Delaunay Triangulation, and therefore GG $\subseteq$ DT.

The 2D Delaunay Triangulation can be constructed in $\Theta(N_v \log N_v)$ time, which is optimal [Lee and Schachter, 80]. According to [Brown, 79], the kD Delaunay Triangulation can be constructed by computing a particular $(k+1)$D Convex Hull, which results in a time complexity of $\mathcal{O}(N_v^{\lceil k/2 \rceil})$, see Section 2.2.8. Alternatively, denoting the number of $(k-1)$-simplices with N_s, the 2D Delaunay Triangulation can be constructed in $\Theta(N_s \log N_s) = \Theta(N_e \log N_e)$ and the kD Delaunay Triangulation in $\mathcal{O}(N_s \log N_s)$ time (due to [Seidel, 86]).

The 2D Delaunay Triangulation is used in very many applications because of its property that it is the triangulation among all possible triangulations of the same vertex set that maximizes the minimum interior angle of all the triangles. For example, for the construction of a piecewise functional surface over a triangulated domain such a property of the triangulation is desirable in order to avoid numerical problems caused by thin triangles [Lawson, 77], [Nielson and Franke, 83]. The Delaunay Triangulation is even part of the definition of the surface interpolant developed by [Farin, 90b]. In any tetrahedralization, each tetrahedron has planar angles (between two edges of a face), dihedral angles (between two faces), and trihedral, or solid angles (between three faces). It is not known if any of the minima of these angles is maximized in the Delaunay Triangulation.

Figure 2.1 shows examples of the geometric graphs mentioned so far.

2.2.10 α-Shape

The α-Shape and α-Hull were introduced by [Edelsbrunner et al., 83] for vertices in 2D, but they are straightforwardly generalized to higher dimensions. The definition of the α-Hull is based on the notion of a parameterized generalized ball:

DEFINITION 2.12 (GENERALIZED BALL) *Let a be an arbitrary real number. A generalized ball of radius $1/a$ is defined as a ball of radius $1/a$ if $a > 0$, the complement of a ball of radius $1/(-a)$ if $a < 0$, and a half-space if $a = 0$.*

DEFINITION 2.13 ($\alpha(a)$-HULL) *Let V be a set of vertices in kD. The $\alpha(a)$-Hull of V is the closure of the intersection of all generalized balls of radius $1/a$ that contain V.*

The $\alpha(a)$-Hull is a bounded closed point set in kD space whose boundary consists of spherical segments of curvature a. The vertices on the boundary of an $\alpha(a)$-Hull are called the extreme vertices. Informally speaking, replacing the circular arcs by line segments (2D), and spherical segments by triangles (3D) gives a (hyper-)graph, called the α-Shape. This is formalized in the following definition:

DEFINITION 2.14 ($\alpha(a)$-SHAPE) *The $\alpha(a)$-Shape of a set of vertices V is the hyper-graph (V, E) with E the set of $(k-1)$-simplices joining k extreme vertices lying on a spherical segment of the boundary of the $\alpha(a)$-Hull that contains no other extreme vertices.*

The $\alpha(a)$-Shape is a subgraph of the closest point Delaunay Triangulation if $a \geq 0$, and a subgraph of the furthest-point Delaunay Triangulation if $a \leq 0$ (the $\alpha(0)$-Shape coincides with the Convex Hull, which is a subgraph of both the closest and the furthest point Delaunay Triangulation). The time complexities for constructing the Delaunay Triangulation therefore carry over to the $\alpha(a)$-Shape.

2.2.11 β-Skeleton

The β-Skeleton is a parameterized neighborhood graph, introduced for 2D in a circle-based and a lune-based variant [Kirkpatrick and Radke, 85]. I will denote them with β_c- and β_l-Skeleton respectively, or $\beta_c(b)$- and $\beta_l(b)$-Skeleton for the specific parameter value b. The corresponding neighborhoods $N_{\beta_c}(b)$ and $N_{\beta_l}(b)$ are defined below. The definitions are a slightly modified version of the originals, in order to normalize the parameter so as to lie between -1 and 1.

DEFINITION 2.15 (CIRCLE-BASED β-NEIGHBORHOOD $N_{\beta_c}(b)$) *Let v_1 and v_2 be two distinct vertices in the plane, and $b \in [-1, 1]$. The $N_{\beta_c}(b)$ of v_1 and v_2 is defined by two discs D_1, D_2 touching v_1 and v_2 with radius $d(v_1, v_2)/2(1 - |b|)$ such that*

if $b \neq 0$:   the centers of D_1, D_2 lie at opposite sides of $v_1 v_2$,
if $b \in [-1, 0]$: $N_{\beta_c}(b) =$ the interior of $(D_1 \cap D_2) \cup$ interior of $v_1 v_2$,
if $b \in [0, 1]$: $N_{\beta_c}(b) =$ the interior of $D_1 \cup$ interior of $D_2 \cup$ interior of $v_1 v_2$.

DEFINITION 2.16 (LUNE-BASED β-NEIGHBORHOOD $N_{\beta_l}(b)$) *Let v_1 and v_2 be two distinct vertices in the plane, and $b \in [-1, 1]$. The $N_{\beta_l}(b)$ of v_1 and v_2 is*

if $b \in [-1, 0]$: $N_{\beta_c}(b)$,
if $b \in [0, 1]$:   the interior of the intersection of the discs that are centered at $v_1 + (v_2 - v_1)/2(1 - b)$ and $v_2 + (v_1 - v_2)/2(1 - b)$, both having radius $d(v_1, v_2)/2(1 - b)$.

Note that the open line segment $v_1 v_2$ is always part of the neighborhood.

The β-Skeleton is a neighborhood graph joining vertices if and only if the associated β-neighborhood is empty. For special values of the parameter the β_l-Skeletons reduce to particular geometric graphs:

- $N_{\beta_l}(0)$ is equivalent to the Gabriel neighborhood. The $\beta_c(0)$- and $\beta_l(0)$-Skeletons are the Gabriel Graph.
- $N_{\beta_l}(\frac{1}{2})$ reduces to the relative neighborhood. The $\beta_l(\frac{1}{2})$-Skeleton is the Relative Neighborhood Graph.
- The $\beta_l(1)$-Skeleton reduces to ∞-SG.
- The $\beta_c(-1)$- and $\beta_l(-1)$-Skeleton are complete graphs if no three vertices are collinear.
- The $\beta_c(1)$-Skeleton is the empty graph.

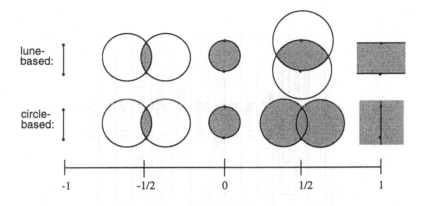

Figure 2.2. Overview of the spectrum of β-neighborhoods in 2D.

The spectrum of β-neighborhoods for the whole range of the parameter is illustrated in Figure 2.2.

The generalization of $N_{\beta_l}(b)$ to higher dimensions is straightforward for $b \geq 0$ (replace the word disc by ball in the definition), but a higher-dimensional $N_{\beta_c}(b)$ could be defined in various ways. The β-Skeleton can be used for the analysis of networks (e.g. the major road network in Saskatchewan, Canada, see [Kirkpatrick and Radke, 85]).

For $b < 0$ the β-Skeletons are equal and can be constructed by a naive brute force algorithm in $\mathcal{O}(N_v^3)$ time. The $\beta_r(b)$-Skeleton for $b \geq 0$ can be constructed in $\mathcal{O}(N_v \log N_v)$ time, and the $\beta_l(b)$-Skeleton, $b \geq 0$, in $\mathcal{O}(N_v^2)$ time, see [Kirkpatrick and Radke, 85].

2.3   Concluding remarks

The inclusion relations between all geometric graphs mentioned in the previous section are depicted in Figure 2.3, in which α represents the α-Shape, and β_c and β_l represent the β-Skeletons. Table 2.1 lists the best known upper bounds of the time complexities to construct the graphs, and the references where these results can be found.

In [Kirkpatrick and Radke, 85] it is said that a geometric graph describes the internal structure of a set of vertices, when it joins essential neighbors among the essential vertices. The external structure is described when the graph joins essential neighbors among the essential extreme vertices. Exactly when vertices or pairs of vertices are considered essential generally depends on the application, and when vertices are considered neighbors depend on the definition of the graph or, when appropriate, the neighborhood. The Convex Hull and its parameterized generalization, the α-Shape, describe aspects of the external structure of a set of vertices. All other geometric graphs mentioned here describe different aspects of the internal structure.

The next chapter introduces the γ-neighborhood graph, which extends the

	2D		$kD, k \geq 3$	
	upper bound	reference	upper bound	reference
CP	$\Theta(N_v \log N_v)$	[Bentley and Shamos, 76]	$\Theta(N_v \log N_v)$	[Bentley and Shamos, 76]
NNG	$\Theta(N_v \log N_v)$	[Bentley and Shamos, 76]	$\mathcal{O}(N_v(\log N_v)^{k-1})$	[Bentley and Shamos, 76]
EMST	$\Theta(N_v \log N_v)$	[Shamos and Hoey, 75]	$\mathcal{O}(N_v^2)$	[Prim, 57]
∞-SG	$\mathcal{O}(N_v^2)$	trivial	$\mathcal{O}(N_v^2)$	trivial
SIG	$\Theta(N_v \log N_v)$	[Toussaint, 88b]	$\mathcal{O}(N_v^2)$	trivial
RNG	$\Theta(N_v \log N_v)$	[Supowit, 83]	$\mathcal{O}(N_v^3)$	[Toussaint, 80]
GG	$\Theta(N_v \log N_v)$	[Matula and Sokal, 80]	$\mathcal{O}(N_v^3)$	trivial
CH	$\Theta(N_v \log N_v)$	[Preparata and Hong, 77]	$\mathcal{O}(N_v \log N_v + N_v^{\lfloor k/2 \rfloor})$	[Chazelle, 91]
DT	$\Theta(N_v \log N_v)$	[Lee and Schachter, 80]	$\mathcal{O}(N_v^{\lceil k/2 \rceil})$	[Brown, 79]+[Chazelle, 91]
α-Shape	$\Theta(N_v \log N_v)$	[Edelsbrunner et al., 83]	$\mathcal{O}(N_v^{\lceil k/2 \rceil})$	[Edelsbrunner et al., 83]
$\beta_c(b)$-, $\beta_l(b)$-Sk., $b < 0$	$\mathcal{O}(N_v^3)$	[Kirkpatrick and Radke, 85]	$\mathcal{O}(N_v^3)$	trivial
$\beta_c(b)$-Skel., $b \geq 0$	$\mathcal{O}(N_v \log N_v)$	[Kirkpatrick and Radke, 85]	not defined	
$\beta_l(b)$-Skel., $b \geq 0$	$\mathcal{O}(N_v^2)$	[Kirkpatrick and Radke, 85]	$\mathcal{O}(N_v^3)$	trivial

Table 2.1. Time complexities for graph construction.

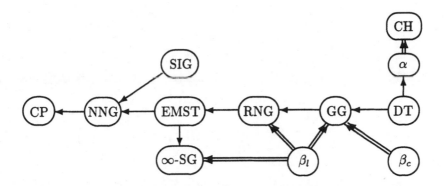

Figure 2.3. Hierarchy of geometric graphs. Graph1 ← graph2 denotes graph1 ⊆ graph2, and graph1 ⊨ graph2 indicates that the parameterized graph2 reduces to graph1 for specific parameter values.

hierarchy in Figure 2.3. The γ-neighborhood graph can describe both internal and external structures of a set of vertices.

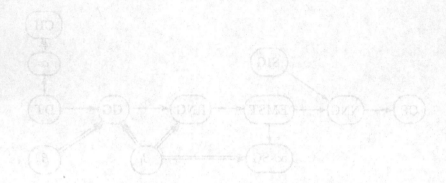

Figure 2.7. Hierarchy of geometric graphs. Graph1 → graph2 denotes graph1 ⊆ graph2 and graph1 = graph2 indicates that the parameterized graph2 reduces to graph1 for specific parameter values.

hierarchy in Figure 2.5. The ε-neighborhood graph can describe both internal and external structures of a set of particles.

3

The γ-Neighborhood Graph

This chapter introduces the γ-Neighborhood Graph, a novel two-parameter geometric graph. It unifies a number of geometric graphs such as the Convex Hull, the Delaunay Triangulation, the Gabriel Graph and the β_c-Skeleton, into a continuous spectrum of geometric graphs that ranges from the empty to the complete graph.

3.1 Introduction

In the computational geometry discipline many old and new geometric techniques are brought together and unified. An example of this is the development in geometric graphs. A major unifying effect in computational geometry was brought about by the Delaunay Triangulation and its dual Voronoi Diagram. Old geometric graphs such as the Convex Hull and the Euclidean Minimum Spanning Tree, and new, parameterized graphs such as the α-Shape and the β_c-Skeleton are intimately related to the Delaunay Triangulation. An even more general graph is presented in this chapter: the γ-Neighborhood Graph.

The Delaunay Triangulation can be seen as a neighborhood graph, where the neighborhood consists of the union of two balls whose radii need not be the same and may have any size. On the other hand, $N_{\beta_c}(b)$ consists of the union or intersection of two discs whose radii must be the same, and are controlled by b. The γ-Neighborhood Graph is based on the combination of these properties of the two graphs.

3.2 The $\gamma(c_0, c_1)$-Graph

The γ-Neighborhood Graph will be used in Chapter 5 in 2D and 3D only; however, for reasons of generality it will here be defined for arbitrary dimension. In the definition of the γ-Neighborhood the following notation is used: for k distinct vertices $v_1, \ldots, v_k$ in kD not lying in a $(k-2)$D plane, $r(v_1, \ldots, v_k)$ denotes the radius of the smallest kD ball touching these vertices, i.e. the radius of the unique $(k-1)$D ball touching the vertices. A two-parameter γ-Neighborhood $N_\gamma(c_0, c_1)$ is defined as follows:

DEFINITION 3.1 (γ-NEIGHBORHOOD $N_\gamma(c_0, c_1)$) *Let $v_1, \ldots, v_k$ be k distinct vertices in kD not lying in a $(k-2)$D hyper-plane, R the closed part of the hyper-plane through $v_1, \ldots, v_k$ bounded by the $(k-1)$-simplex $v_1 \ldots v_k$, and $c_0, c_1 \in [-1, 1]$ such that $|c_0| \le |c_1|$. A $N_\gamma(c_0, c_1)$ is defined by the kD balls B_0, B_1 of radius $r(v_1, \ldots, v_k)/(1 - |c_0|)$ and $r(v_1, \ldots, v_k)/(1 - |c_1|)$ respectively and touching $v_1, \ldots, v_k$, such that*

if $c_0 c_1 < 0$:   *the centers of B_0, B_1 lie on the same side of the hyper-plane through $v_1, \ldots, v_k$,*

if $c_0 c_1 > 0$:   *the centers of B_0, B_1 lie on opposite sides of the hyper-plane through $v_1, \ldots, v_k$,*

if $c_1 \in [-1, 0]$: $N_\gamma(c_0, c_1) =$ the interior of $(B_0 \cap B_1) \cup$ interior of R,

if $c_1 \in [0, 1]$: $N_\gamma(c_0, c_1) =$ the interior of $B_0 \cup$ interior of $B_1 \cup$ interior of R.

Note that the interior of R, the open simplex $v_1 \ldots v_k$, is always part of the neighborhood.

Figure 3.1 gives a graphical overview of the whole spectrum of 2D neighborhoods. Note that the definition is valid for $c_1 = 0$: since $|c_0| \le |c_1|$, c_0 must also be zero, the radii of both balls are $r(v_1, \ldots, v_k)$, so that the balls have the same center and the intersection gives the same result as the union. For given $v_1, \ldots, v_k$ and any $c_0, c_1 \in [-1, 1]$ such that $|c_0| < |c_1|$ (so $c_0 \ne c_1$), there are two $N_\gamma(c_0, c_1)$'s, which are mirror-symmetric with respect to the plane through $v_1 \ldots v_k$. The γ-Graph $\gamma(c_0, c_1)$ joins the vertices if at least one of the neighborhoods is empty:

DEFINITION 3.2 ($\gamma(c_0, c_1)$ NEIGHBORHOOD GRAPH) *Let V be a set of vertices in kD. The $\gamma(c_0, c_1)$-Graph is the hyper-graph (V, S) with S the set of $(k-1)$-simplices $v_1 \ldots v_k$ such that the $N_\gamma(c_0, c_1)$ of $v_1, \ldots, v_k$ is empty.*

I will use 'γ-Graph', '$\gamma(c_0, c_1)$-Graph' or simply '$\gamma(c_0, c_1)$' and similar expressions, to denote the appropriate γ-Neighborhood Graph.

In a 2D γ-Graph (V, E) the elements of E are edges, joining pairs of vertices. Associated with an edge $v_1 v_2$ is a $N_\gamma(c_0, c_1)$ defined by two discs whose radii are a scaling factor times the radius of the smallest circle through $v_1 v_2$, i.e. $r(v_1, v_2) = d(v_1, v_2)/2$. For special values of c_0 and c_1 the 2D $\gamma(c_0, c_1)$ reduces to particular geometric graphs:

- $c_0 = c_1 = 1$. The neighborhood is the entire plane except for two half-lines originating at v_1 and v_2. If no three vertices are collinear, $\gamma(1, 1)$ is an empty graph.
- $c_0 = c_1 = -1$. The neighborhood is the line segment $v_1 v_2$. If no three vertices are collinear, then $\gamma(-1, -1)$ is the complete graph.

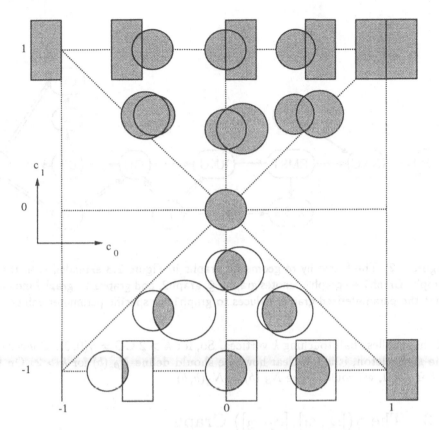

Figure 3.1. Overview of the spectrum of 2D neighborhoods $N_\gamma(c_0, c_1)$. The neighborhoods drawn are not appropriately scaled, but give an idea of their shape. Rectangles denote half-spaces.

- $c_0 = -1$, $c_1 = 1$ and $c_0 = 1$, $c_1 = -1$. In both cases the two half-planes lie
  on the same side of the line through v_1 and v_2 and coincide (more generally,
  $N_\gamma(c_0, -c_0) = N_\gamma(-c_0, c_0)$). The neighborhood is empty if all other vertices
  lie on one side of the line through v_1 and v_2, or *on* the line except on line
  segment $v_1 v_2$. That occurs if and only if v_1 and v_2 lie on the Convex Hull.
  Therefore, $\gamma(-1, 1)$ and $\gamma(1, -1)$ are the Convex Hull.
- $c_0 = c_1$. $N_\gamma(c_0, c_0) = N_{\beta_c}(c_0)$. So, $\gamma(c_0, c_0)$ reduces to the β_c-Skeleton.
- $c_0 = c_1 = 0$. $N_\gamma(0,0)$ is the smallest disc touching v_1 and v_2, which is the
  Gabriel Neighborhood. So, $\gamma(0,0)$ is the Gabriel Graph.

In a 3D γ-Graph (V, T), the elements of T are triangles joining three vertices.
Associated with a triangle $v_1 v_2 v_3$ is a $N_\gamma(c_0, c_1)$ defined by two balls, whose radii
are a scaling factor times the radius of the smallest sphere through v_1, v_2, and
v_3, i.e. $r(v_1, v_2, v_3)$.

Again for special values of c_0 and c_1, the kD $\gamma(c_0, c_1)$, $k \geq 3$, reduces to
particular graphs: $\gamma(1, 1)$ is the empty graph, $\gamma(-1, -1)$ is the complete graph,
and $\gamma(-1, 1) = \gamma(1, -1) = \text{CH}$. Note that the kD Gabriel Neighborhood is the
interior of the smallest ball touching two vertices, whereas $N_\gamma(0, 0)$ is the interior

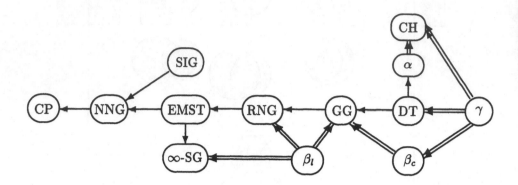

Figure 3.2. The hierarchy of geometric graphs in Figure 2.3 extended with the γ-Graph. Graph1 ← graph2 denotes graph1 ⊆ graph2, and graph1 ⇐ graph2 indicates that the parameterized graph2 reduces to graph1 for specific parameter values.

of the smallest ball touching k vertices. So, for $k > 2$ GG $\neq \gamma(0,0)$. Concerning the β_c-Skeleton, it is not clear how one should define $N_{\beta_c}(b)$ for $k > 2$. On the other hand, we could define $N_{\beta_c}(b) = N_\gamma(b,b)$.

3.3 The $\gamma([c_0,c_1],[c_2,c_3])$-Graph

So far the γ-parameters were fixed. We can also consider the largest values of the γ-parameters, for which the corresponding neighborhood is still empty. That is the value for which the ball touches a $(k+1)$th vertex, or is either 1 or -1 if there is no such vertex. This concept is provided by the $\gamma([c_0,c_1],[c_2,c_3])$-Graph, defined as follows:

DEFINITION 3.3 ($\gamma([c_0,c_1],[c_2,c_3])$ NEIGHBORHOOD GRAPH) *Let V be a set of vertices in kD. $\gamma([c_0,c_1],[c_2,c_3])$ is the hyper-graph (V,S) with S the set of $(k-1)$-simplices $v_1 \ldots v_k$ for which the largest values of c_0', c_1' such that the associated $N_\gamma(c_0',c_1')$ is empty satisfy $c_0' \in [c_0,c_1]$ and $c_1' \in [c_2,c_3]$.*

The $\gamma([-1,1],[0,1])$-Graph joins vertices $v_1,\ldots,v_k$ in kD if there are two empty balls of arbitrary radius touching the vertices. This is exactly a definition of the Delaunay Triangulation if no more than $k+1$ vertices lie on an empty ball, see Definition 2.11 on page 15. If more than $k+1$ vertices lie on an empty ball, $\gamma([-1,1],[0,1])$ forms all possible overlapping $(k-1)$-simplices, whereas the degenerate Delaunay Triangulation arbitrarily forms as much as possible non-overlapping $(k-1)$-simplices.

The relation between the 2D γ-Graph and the 2D geometric graphs mentioned in Section 2.2 is depicted in Figure 3.2.

The γ-Graph not only describes the internal structure of a set of vertices, but also aspects of the external structure. For example, the $\gamma(-1,1)$-Graph reduces to the Convex Hull. The next section gives an example in which special γ-parameter values give a clear external structure. We will see in Chapter 5 how the γ-Graph is used to find a boundary through all vertices. This capability of describing the external structure somewhat contrasts the β_l-Skeleton. N_{β_l} is located inside the infinite strip (Section 2.2.4) of the two vertices (2D). The resulting graph therefore emphasizes connections between vertices, which makes it suitable for network analysis. The two balls in N_γ are located aside the k vertices involved (kD). Especially when N_γ is the union of the balls, the γ-Graph is more like (a part of) a tessellation.

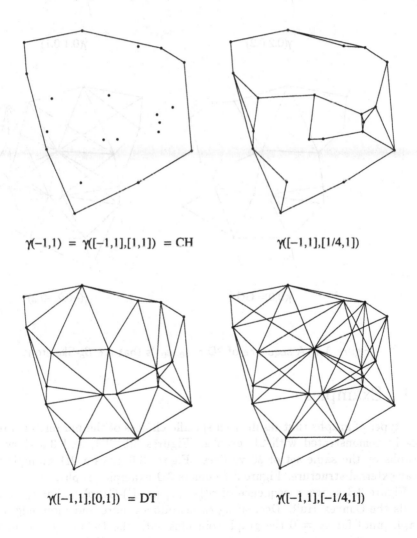

$$\gamma(-1,1) \;=\; \gamma([-1,1],[1,1]) \;=\; CH \qquad\qquad \gamma([-1,1],[1/4,1])$$

$$\gamma([-1,1],[0,1]) \;=\; DT \qquad\qquad\qquad \gamma([-1,1],[-1/4,1])$$

Figure 3.3. A sequence of 2D $\gamma([-1,1], [c_0, 1])$'s, successively containing more edges.

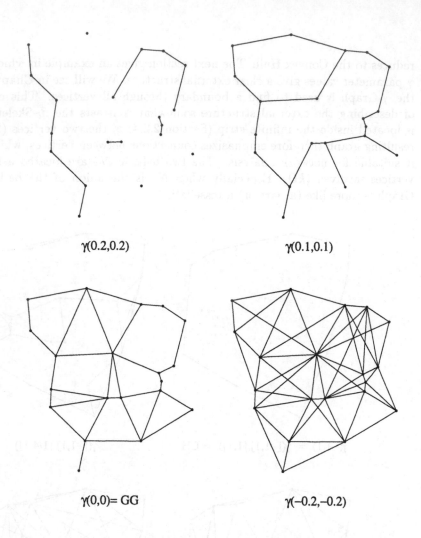

$\gamma(0.2,0.2)$ $\gamma(0.1,0.1)$

$\gamma(0,0)=$ GG $\gamma(-0.2,-0.2)$

Figure 3.4. A sequence of 2D $\gamma(c_0, c_0)$'s that are β_c-Skeletons.

3.4   Examples

The types of graphs that result from specific choices of the parameters are most clearly demonstrated with 2D graphs. Figures 3.3, 3.4, and 3.5 show 2D γ-Graphs on the same set of 20 vertices. Figure 3.6 gives a 2D example with a clear external structure. Figure 3.7 depicts 3D example graphs.

Figure 3.3 shows a sequence of $\gamma([-1,1], [c_0,1])$-Graphs. For $c_0 = 1$, this yields the Convex Hull. Decreasing c_0 introduces more and more edges in the graph, until for $c_0 = 0$ the graph coincides with the Delaunay Triangulation. When c_0 gets negative, edges may cross each other, until for $c_0 = -1$ the graph would be complete (not shown).

The graphs in Figure 3.4 all coincide with a β_c-Skeleton. The neighborhoods $N_\gamma(0.2,0.2)$, $N_\gamma(0.1,0.1)$, $N_\gamma(0,0)$, and $N_\gamma(-0.2,-0.2)$ get successively

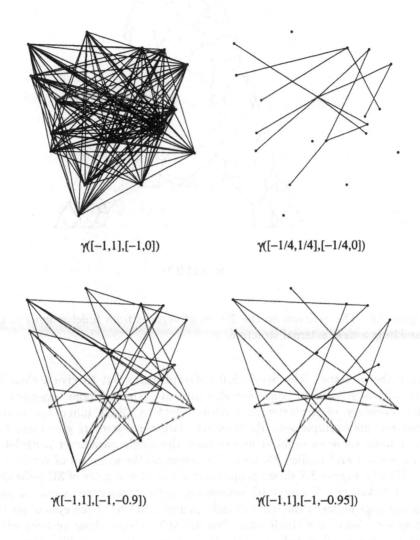

Figure 3.5. A sequence of 2D γ-Graphs where the neighborhoods consist of intersections of discs.

smaller. The emptiness requirement gets less restrictive, so that more pairs of vertices are considered neighbors. The $\gamma(0,0)$-Graph equals the Gabriel Graph.

Figure 3.5 depicts graphs that result when only intersections of discs are allowed as neighborhood. The $\gamma([-1,1],[-1,0])$ joins all pairs of vertices that have no empty ball. It it the complement of the Delaunay Triangulation. In $\gamma([-\frac{1}{4},\frac{1}{4}],[-\frac{1}{4},0])$, the intersections are forced to have a certain minimal width. On the other hand, $\gamma([-1,1],[-1,-0.9])$ allows only thin neighborhoods. The edges then join vertices only when there is another vertex close to the edge. In $\gamma([-1,1],[-1,-0.95])$ the neighborhoods are so thin, that vertices are only joined if there is another vertex almost on the edge.

Figure 3.6 shows the set of vertices used in [Edelsbrunner et al., 83] to illus-

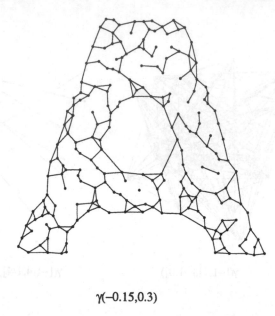

$\gamma(-0.15, 0.3)$

Figure 3.6. The γ-Graph on the 2D set of vertices from [Edelsbrunner et al., 83], exhibiting a clear external structure.

trate the α-Shape. The $\gamma(-0.15, 0.3)$-Graph turns out to give a clear boundary, although internal edges are also present. The α-Shape, designed to give the boundary of a cluster of vertices, yields a single inner and outer contour on this example set. However, the two γ-parameters give more freedom for finding some external structure than the single parameter β-Skeleton; see [Kirkpatrick and Radke, 85] for a β-Skeleton on the same set of vertices.

Finally, Figure 3.7 shows projections of two stereo pairs of 3D γ-Graphs of a set of 30 vertices. To get a 3D impression, use a stereoscope or place a partition on the page between the left and right picture, and relax the eyes to let the two images coalesce to a single one. The 3D $\gamma(0,0)$ joins three vertices with each other if the smallest ball touching these vertices is empty. The picture of the $\gamma(0,0)$ only slightly differs from a typical 3D Delaunay Triangulation. This is because three pairwise incident edges of three different triangles in the graph can give the false impression of forming a fourth triangle. In the example shown, the $\gamma(0,0)$ consists of 150 triangles, while the Delaunay Triangulation on the same set of vertices consists of 257 triangles (constituting 248 tetrahedra). The $\gamma([-\frac{1}{4}, \frac{1}{4}], [\frac{1}{2}, \frac{3}{4}])$ is disconnected, but shows more clearly that the 3D γ-Graph consists of triangles.

3.5   Complexities

The following three lemmas tell how γ-Graphs with specific parameter values are related to each other. The lemmas give cues how to construct an arbitrary γ-Graph.

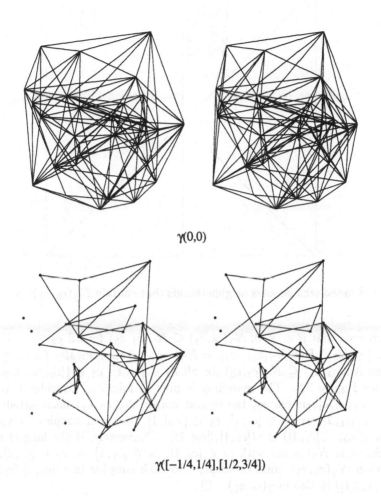

$\gamma(0,0)$

$\gamma([-1/4,1/4],[1/2,3/4])$

Figure 3.7. Two stereo pairs of perspectively projected 3D γ-Graphs.

LEMMA 3.1 *Let* $c_0 \in [-1, 1]$, $c_1 \in [0, 1]$, *and* $|c_0| \leq |c_1|$. *Then* $\gamma(c_0, c_1)$ *is equal to* $\gamma([c_0, 1], [c_1, 1])$.

Proof. Observe that if $c_1 \geq 0$, then $N_\gamma(c_0, c_1) \subseteq N_\gamma(c_2, c_3)$ for all $c_2 \in [c_0, 1]$, $c_3 \in [c_1, 1]$, $|c_2| \leq |c_3|$, see Figure 3.8. The reasoning is now as follows. Consider k vertices in kD. If $N_\gamma(c_0, c_1)$ is empty, then the largest empty $N_\gamma(c_2, c_3)$ must satisfy $c_2 \in [c_0, 1]$, and $c_3 \in [c_1, 1]$. So each simplex in $\gamma(c_0, c_1)$ is also in $\gamma([c_0, 1], [c_1, 1])$. Conversely, if the largest empty γ-Neighborhood is $N_\gamma(c_2, c_3)$ with $c_2 \in [c_0, 1]$, and $c_3 \in [c_1, 1]$, then $N_\gamma(c_0, c_1)$ must be empty. So each simplex in $\gamma([c_0, 1], [c_1, 1])$ is also in $\gamma(c_0, c_1)$. $\square$

LEMMA 3.2 *Let* $c_0 \in [-1, 1]$, $c_1 \in [-1, 0]$, *and* $|c_0| \leq |c_1|$. *Then* $\gamma(c_0, c_1)$ *is equal to* $\gamma([c_0, 1], [c_1, 1]) \cup \gamma([c_1, 1], [|c_0|, 1])$.

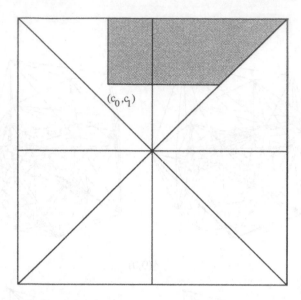

Figure 3.8. Shaded area denotes neighborhoods that contain $N_\gamma(c_0, c_1)$, $c_1 \in [0, 1]$.

Proof. Observe that if $c_1 \leq 0$, $N_\gamma(c_0, c_1) \subseteq N_\gamma(c_2, c_3)$ for all $c_2 \in [c_0, 1]$, $c_3 \in [c_1, 1]$, $|c_2| \leq |c_3|$. But $N_\gamma(c_4, -c_4) = N_\gamma(-c_4, c_4)$, specifically for $c_1 \leq c_4 \leq c_0$, and so $N_\gamma(c_0, c_1) \subseteq N_\gamma(c_2, c_3)$ for all $c_2 \in [c_1, 1]$, $c_3 \in [|c_0|, 1]$, $|c_2| \leq c_3$ as well, see Figure 3.9. The reasoning is now as follows. Consider k vertices in kD. If $N_\gamma(c_0, c_1)$ is empty, the largest empty $N_\gamma(c_2, c_3)$ must satisfy $c_2 \in [c_0, 1]$, $c_3 \in [c_1, 1]$, or $c_2 \in [c_1, 1]$, $c_3 \in [|c_0|, 1]$. So each simplex in $\gamma(c_0, c_1)$ is also in $\gamma([c_0, 1], [c_1, 1]) \cup \gamma([c_1, 1], [|c_0|, 1])$. Conversely, if the largest empty neighborhood is $N_\gamma(c_2, c_3)$ with $c_2 \in [c_0, 1]$, $c_3 \in [c_1, 1]$, or $c_2 \in [c_1, 1]$, $c_3 \in [|c_0|, 1]$, then $N_\gamma(c_0, c_1)$ must be empty. So each simplex in $\gamma([c_0, 1], [c_1, 1]) \cup \gamma([c_1, 1], [|c_0|, 1])$ is also in $\gamma(c_0, c_1)$. $\square$

LEMMA 3.3 $\gamma([c_0, c_1], [c_2, c_3]) \subseteq \gamma([c_4, c_5], [c_6, c_7])$ *on all sets of vertices if and only if* $[c_0, c_1] \subseteq [c_4, c_5]$ *and* $[c_2, c_3] \subseteq [c_6, c_7]$.

Proof. Let $[c_0, c_1] \subseteq [c_4, c_5]$ and $[c_2, c_3] \subseteq [c_6, c_7]$, and consider a simplex in the graph $\gamma([c_0, c_1], [c_2, c_3])$. Its largest empty $N_\gamma(c_0', c_1')$ satisfies $c_0' \in [c_0, c_1] \subseteq [c_4, c_5]$, and $c_1' \in [c_2, c_3] \subseteq [c_6, c_7]$. That simplex is thus also present in $\gamma([c_4, c_5], [c_6, c_7])$. Conversely, let $\gamma([c_0, c_1], [c_2, c_3]) \subseteq \gamma([c_4, c_5], [c_6, c_7])$. Every simplex in $\gamma([c_0, c_1], [c_2, c_3])$ is also in $\gamma([c_4, c_5], [c_6, c_7])$ and has a largest empty $N_\gamma(c_0', c_1')$ that satisfies $c_0' \in [c_0, c_1]$ and $c_0' \in [c_4, c_5]$, and also $c_1' \in [c_2, c_3]$ and $c_1' \in [c_6, c_7]$. Therefore $[c_0, c_1] \subseteq [c_4, c_5]$ and $[c_2, c_3] \subseteq [c_6, c_7]$. $\square$

Lemmas 3.1, 3.2, and 3.3 are illustrated in Figure 3.10 by means of the 2D graphs $\gamma(\frac{1}{4}, \frac{1}{2})$, $\gamma(-\frac{1}{4}, \frac{1}{4})$, $\gamma(0, -\frac{1}{4})$, and $\gamma(0, -\frac{1}{2})$. By Lemma 3.1, $\gamma(\frac{1}{4}, \frac{1}{2}) = \gamma([\frac{1}{4}, 1], [\frac{1}{2}, 1])$ (say G1), and $\gamma(-\frac{1}{4}, \frac{1}{4}) = \gamma([-\frac{1}{4}, 1], [\frac{1}{4}, 1])$ (G2). Lemma 3.2 tells $\gamma(0, -\frac{1}{4}) = \gamma([0, 1], [-\frac{1}{4}, 1]) \cup \gamma([-\frac{1}{4}, 1], [0, 1])$ (G3), and $\gamma(0, -\frac{1}{2}) =$

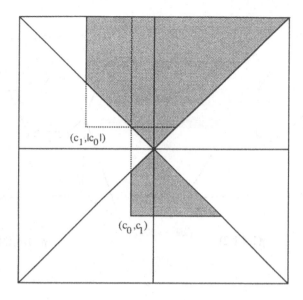

Figure 3.9. Shaded area denotes neighborhoods that contain $N_\gamma(c_0, c_1)$, $c_1 \in [-1, 0]$.

$\gamma([0, 1], [-\frac{1}{2}, 1]) \cup \gamma([-\frac{1}{2}, 1], [0, 1])$ (G4). According to Lemma 3.3, G1 $\subseteq$ G2 $\subseteq$ G3 $\subseteq$ G4, which is illustrated in the figure.

The preceding lemmas can be used to derive necessary time complexities for construction of γ-Graphs. By Lemma 3.1 and 3.2, any $\gamma(c_4, c_5)$ can be expressed in terms of graphs of the form $\gamma([c_0, c_1], [c_2, c_3])$, so that only the latter type of graphs need to be considered. Time and storage complexities are directly related to the size of the graph.

The size of a γ-Graph (V, S) is the number of simplices in S, say M. So in 2D $M = N_e$ and in 3D $M = N_t$. For a fixed dimension k, the amount of space to store a simplex is constant, so the total storage complexity is $\Theta(M)$. Naturally, the necessary time for construction by any algorithm is at least linear in M: $\Omega(M)$. For an arbitrary γ-Graph M is given by the following lemma:

LEMMA 3.4 *Let V be a set of vertices in kD. The number of $(k-1)$-simplices in an arbitrary $\gamma([c_0, c_1], [c_2, c_3])$ is bounded by $\mathcal{O}(N_v^k)$.*

Proof. The upper bound is equal to the number of $(k-1)$-simplices in the complete γ-Graph, which is $\mathcal{O}(\binom{N_v}{k}) = \mathcal{O}(N_v^k)$. □

In the restricted, but still large, class of cases that $[c_0, c_1] \subseteq [-1, 1]$ and $[c_2, c_3] \subseteq [0, 1]$, $\gamma([c_0, c_1], [c_2, c_3])$ consists of much less simplices than stated by Lemma 3.4, provided that the set of vertices is non-degenerate, i.e. that no empty ball touches $k + 2$ vertices.

LEMMA 3.5 *Let V be a non-degenerate set of vertices in kD, and let $[c_0, c_1] \subseteq [-1, 1]$ and $[c_2, c_3] \subseteq [0, 1]$. The number of $(k-1)$-simplices in $\gamma([c_0, c_1], [c_2, c_3])$ is $\mathcal{O}(N_v^{\lceil k/2 \rceil} / \lceil k/2 \rceil!)$.*

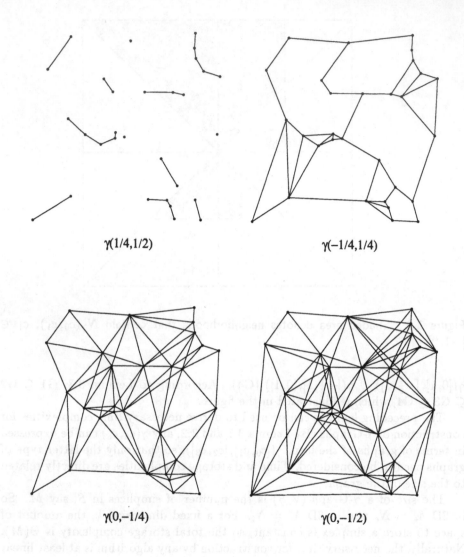

$\gamma(1/4,1/2)$ $\gamma(-1/4,1/4)$

$\gamma(0,-1/4)$ $\gamma(0,-1/2)$

Figure 3.10. Four γ-Graphs on the same set of 30 vertices. $\gamma(\frac{1}{4},\frac{1}{2}) \subseteq \gamma(-\frac{1}{4},\frac{1}{4}) \subseteq \gamma(0,-\frac{1}{4}) \subseteq \gamma(0,-\frac{1}{2})$.

Proof. We know by Lemma 3.3 that $\gamma([c_0,c_1],[c_2,c_3]) \subseteq \gamma([-1,1],[0,1])$, under the stated conditions. Because V is non-degenerate, $\gamma([-1,1],[0,1]) = \text{DT}$. The number of $(k-1)$-simplices in the Delaunay Triangulation is of the same order as the number of k-simplices in the Delaunay Triangulation, which is equal to the number of vertices in the Voronoi Diagram, see Section 2.2.9. The maximum number of vertices in a $k\text{D}$ Voronoi Diagram is $\Theta(N_v^{\lceil k/2 \rceil}/\lceil k/2 \rceil!)$ [Klee, 80]. Since $\gamma([c_0,c_1],[c_2,c_3]) \subseteq \text{DT}$, this is an upper bound for the size of this γ-Graph. $\square$

THEOREM 3.1 *Let V be a set of vertices in kD. Any $\gamma([c_0, c_1], [c_2, c_3])$ of V can be constructed in $\mathcal{O}(N_v^{k+1})$ time.*

Proof. A brute force algorithm takes all the $\binom{N_v}{k}$ possible combinations of k vertices, considers all the $N_v - k$ other vertices to determine the largest empty $N_\gamma(c_0', c_1')$ in constant time per vertex, and tests whether $c_0' \in [c_0, c_1]$ and $c_1' \in [c_2, c_3]$. This amounts to $\mathcal{O}\left((N_v - k)\binom{N_v}{k}\right) = \mathcal{O}(N_v^{k+1})$ time. $\square$

If $[c_0, c_1] \subseteq [-1, 1]$ and $[c_2, c_3] \subseteq [0, 1]$ and the set of vertices is non-degenerate, $\gamma([c_0, c_1], [c_2, c_3])$ can be constructed much more efficiently than stated by Theorem 3.1.

THEOREM 3.2 *Let V be a non-degenerate set of vertices in kD, and let $[c_0, c_1] \subseteq [-1, 1]$ and $[c_2, c_3] \subseteq [0, 1]$. Then $\gamma([c_0, c_1], [c_2, c_3])$ can be computed in a time complexity of $\mathcal{O}(N_v^{\lceil k/2 \rceil})$.*

Proof. We know by Lemma 3.3 that $\gamma([c_0, c_1], [c_2, c_3]) \subseteq \gamma([-1, 1], [0, 1])$, under the stated conditions. Because V is non-degenerate, $\gamma([-1, 1], [0, 1]) = $ DT. After constructing the Delaunay Triangulation, the largest empty $N_\gamma(c_0', c_1')$ of each $(k - 1)$-simplex can be determined in constant time. The test whether $c_0' \in [c_0, c_1]$, $c_1' \in [c_2, c_3]$ also takes constant time per $(k - 1)$-simplex. The total time complexity is therefore bounded by the time to construct the Delaunay Triangulation, which is $\mathcal{O}(N_v^{\lceil k/2 \rceil})$, see Section 2.2.9. $\square$

For the 2D Delaunay Triangulation, $\mathcal{O}(N_v \log N_v)$ is optimal. Whether this is optimal for the γ-Graph depends on the parameter values. It is clearly not optimal when the γ-Graph reduces to the empty graph. The $\mathcal{O}(N_v^{\lceil k/2 \rceil})$ time only applies to non-degenerate cases. For example, in the degenerate case that all vertices lie on an empty kD ball, the graph consists of $\binom{N_v}{k} = \Theta(N_v^k)$ simplices which must all be generated. In that case any correct algorithm must run in $\Omega(N_v^k)$ time.

One may wonder if in an average case the γ-Graph can be constructed in less time than given by the upper bounds in Theorems 3.1 and 3.2. This is true under certain conditions:

THEOREM 3.3 *Let V be a non-degenerate set of vertices uniformly distributed within a kD ball, and let $[c_0, c_1] \subseteq [-1, 1]$ and $[c_2, c_3] \subseteq [0, 1]$. Then $\gamma([c_0, c_1], [c_2, c_3])$ can be constructed in $\mathcal{O}(N_v)$ expected time.*

Proof. Under the stated conditions, $\gamma([c_0, c_1], [c_2, c_3]) \subseteq$ DT. The Delaunay Triangulation can be computed in $\Theta(N_v)$ expected time [Dwyer, 89] (which is optimal), so that $\gamma([c_0, c_1], [c_2, c_3])$ can be constructed in $\mathcal{O}(N_v)$ expected time. $\square$

Again, whether $\mathcal{O}(N_v)$ is optimal depends on the values of the γ-parameters.

3.6   Concluding remarks

The γ-Graph describes the internal structure of a set of vertices, and is capable of describing the external structure for well-chosen parameter values. The inclusion hierarchy in 2D CP $\subseteq$ NNG $\subseteq$ EMST $\subseteq$ RNG $\subseteq$ GG $\subseteq$ DT has been extended: DT $\subseteq$ $\gamma([-1,1],[c_0,1])$, $c_0 \leq 0$. The γ-Graph provides a general framework for describing neighborhood graphs. It unifies the Convex Hull, the Delaunay Triangulation, and in 2D also the Gabriel Graph and the β_c-Skeleton, into a continuous spectrum ranging from the empty to the complete graph.

The neighborhood $N_\gamma(c_0, c_1)$ is only defined for $c_0, c_1 \in [-1, 1]$, and $|c_0| \leq |c_1|$. For k vertices in kD and specific parameters c_0, c_1, there can be two such neighborhoods, which are mirror symmetric in the plane through the k vertices. The parameters $|c_0| \geq |c_1|$ could be used to completely specify the position of the balls. For example, c_0 specifies 'the left', and c_1 'the right' sphere (left and right properly defined with respect to the plane through the k vertices). In this way one could control in which direction the larger ball must lie. However, there is in general no need to specify a preference of direction.

There are several directions for further research. The most urgent is the development of output-sensitive algorithms. We have seen that the γ-Graph can be constructed efficiently if it is a subgraph of the Delaunay Triangulation. Of course for $c_1 < 0$, the size of the $\gamma(c_0, c_1)$-Graph is $\mathcal{O}(N_v^3)$, but an algorithm having a time complexity that depends on the size of the output can probably do better than $\mathcal{O}(N_v^3)$ in most cases. Also for $c_1 > 0$ an output-sensitive algorithm can be profitable, since the size of the $\gamma(c_0, c_1)$-Graph may be sub-linear in N_v. Little is known from stochastic geometry about probabilistic properties of geometric graphs (some results are known about the Delaunay Triangulation [Miles, 70], the Gabriel Graph, and the Relative Neighborhood Graph [Devroye, 88]). Results on the expected number of edges in the γ-Graph may lead to the development of efficient algorithms for the average case.

Another research suggestion is the generalization of γ-Graphs to sets of weighted vertices. The idea behind weighted vertices is that a greater weight has more influence, e.g. in such a way that vertices are connected more easily. Stating the latter more formally, the condition to let vertices be neighbors is less restrictive if their weight is greater. One way to model this is as follows. Let $w_i \in \mathbb{R}, w_i \geq 0$ be the weight of vertex v_i. The k vertices $v_{i_0}, \ldots, v_{i_k}$ in kD are neighbors in the weighted $\gamma(c_0, c_1)$ if no v_j lies inside $N_\gamma(c_0 w_{i_0} \ldots w_{i_k}/w_j, c_1 w_{i_0} \ldots w_{i_k}/w_j)$, for all $j \neq i_0, \ldots, i_k$.

Another generalization is the concept of n-γ-Graphs, analogous to other geometric graphs, like the n-Relative Neighborhood Graph [Su and Chang, 91]. The n-γ-Graph in kD joins k vertices if the associated γ-Neighborhood contains less than n vertices. For both generalizations, the research issues involved are the development of efficient construction algorithms and the examination of applications. For the weighted γ-Graph another research topic is the development of other weighting schemes.

4

Boundary construction

This chapter introduces the problem of constructing a closed object boundary from a set of scattered points, i.e. sets in which no structural relation between the points is known. It is demonstrated that geometric graphs, describing geometrical relations between the points, are useful tools for the construction problem. An overview of existing solutions and their shortcomings is given.

4.1 Introduction

From this chapter on, I shall consider 2D and 3D objects only. In several applications in geometric modeling one needs an unambiguous geometric model, but the initial data is often a set of vertices in 2D or 3D. One such application is product design, where the points are synthetic, and another application is object reconstruction, where the points are measured from the boundary of an existing object. The boundary constructed from the set of points can then for example be used for the initial design of an artifact, for numerical analysis, or for graphical display.

Points on the boundary of an object can be obtained in a variety of ways. If a set of points from a 2D object is given in a sequential order along the boundary, for example obtained by tracing the boundary, the points form a contour chain representing the boundary curve. If a set of points from a 3D object is given by a pile of contours, for example obtained from parallel object cross sections, the points on pairs of consecutive contours can be joined so as to generate ribbons of triangles which in turn form a closed boundary surface. However, there are numerous data sources that do not yield such a clear structural relation between the points in the data set:

- A typical laser range system measures 3D points on the surface of an object by emitting a laser beam at certain x, y-coordinates and inferring the corresponding z-coordinates [Rioux and Cournoyer, 88]. Parts of the surface can be hidden and become visible only after rotating the object or moving around it, because of shadowing by self-occlusion [Corby and Mundy, 90]. A horizontal scan of points will thus generally not yield points that are successively adjacent on the surface. So, the order in which the points are acquired provides no adjacency relation.
- 3D coordinates of boundary points can be computed from stereographic images, for example X-ray images, or from a sequence of images of a moving object [Shirai, 87]. In both cases the set of calculated coordinates generally provides no topological relation between the points.
- Image processing techniques for low level computer vision can extract feature data like object corners from images [Haralick and Shapiro, 92], which can be converted into 2D coordinates. However, information such as the topological relation between the points are not obtained by this low level processing.

If there is no structural relation between the points, or no such information is available because the data source is not known, or if a single boundary construction method is to be used for data from different sources, no structural relation between the points may be assumed. The only a priori knowledge is that they lie on the boundary of an object.

4.2   Statement of the problem

The problem stated so far, "construct an object boundary given a set of boundary points", is ill-stated. To get a better statement we have to make some restrictions. First of all we restrict the constructed object to have no holes. So, a 2D object must be bounded by a single contour, it has no inner contour; the boundary must be topologically equivalent to a circle. A 3D object must be bounded by a single surface and may not have through-passages (like a torus); the boundary must be topologically equivalent to a sphere. Furthermore, we restrict the boundary to be piecewise linear. So, a 2D boundary must consist of line segments and a 3D boundary of flat polygons. Since a triangle is the only n-gon through n vertices in 3D that is guaranteed to be flat, the class of 3D boundaries is further restricted to consist of triangles.

The 2D boundary construction problem is formally stated as follows:

BOUNDARY CONSTRUCTION 2D   *Let V be a set of vertices in 2D. Find a simple polygon through all vertices.*

By definition, a simple polygon is topologically equivalent to a circle, and therefore a valid boundary of an object.

Given a graph (V, E), a simple polygon through all vertices and consisting of edges from E is equivalent to a Hamilton cycle, see page 10. I will occasionally call a Hamilton cycle a Hamilton polygon.

The 3D boundary construction problem is formally stated as follows:

BOUNDARY CONSTRUCTION 3D   *Let V be a set of vertices in 3D. Find a simple closed polyhedron of triangular faces through all vertices.*

By definition, a simple polyhedron is topologically equivalent to a sphere, and thus a valid object boundary. Given a hyper-graph (V, T) with T a set of triangles, I will call a simple polyhedron through all vertices and consisting of faces from T a Hamilton polyhedron. Note that this is not a Hamilton cycle of edges in 3D.

The boundary construction problems are clearly under-constrained, so that a solution is not unique. Indeed, a set of vertices is an ambiguous boundary representation. Some criterion is needed to select the boundary that is considered the best among all solutions. However, there is no known algorithm that generates all solutions efficiently.

A brute force algorithm for 2D, where a boundary must consist of N_v edges, takes all combinations of N_v edges out of all $\binom{N_v}{2}$ possible edges and tests whether each of the combinations is a simple polygon through all vertices. This gives a time complexity of

$$\Theta\left(\binom{\binom{N_v}{2}}{N_v}\right),$$

which is at least $\Omega(N_v^{N_v})$. In 3D, where a boundary must consist of $2N_v - 4$ triangles (see page 11), the analogous algorithm gives a time complexity of

$$\Theta\left(\binom{\binom{N_v}{3}}{2N_v - 4}\right),$$

which is at least $\Omega(N_v^{5N_v})$.

Such brute force approaches are clearly infeasible. It is logical to exploit some geometric relation between the vertices. In particular we can construct a geometric graph on the set of vertices and take advantage of the incorporated geometric property. As a first try one can think of a shortest distance property, as in the Nearest Neighbors Graph. Many edges in the Nearest Neighbors Graph can be part of a valid boundary, but the shortest distance does not always suitably correspond with the metric on the object boundary, specifically at highly curved parts. As a result, nearest neighbors need not be consecutive points on the boundary. This is illustrated in Figure 4.1.

One can also think of other geometric graphs such as the Convex Hull, giving a rough approximation of a boundary, or the ubiquitous Delaunay Triangulation. An advantage of the Delaunay Triangulation is that it is a connected graph that contains not as few joins as the Euclidean Minimum Spanning Tree, and also not as many as the complete graph. In fact, it contains the maximum number of non-crossing $(k-1)$-simplices. However, also a brute force search in the Delaunay Triangulation for a collection of $(k - 1)$-simplices that forms a valid boundary is not feasible. After all, a Delaunay Triangulation in 2D consists of $3N_v - 6$ edges, giving a time complexity of $\Theta(\binom{3N_v - 6}{N_v})$ which is $\Omega(3^{N_v})$. In 3D the maximum

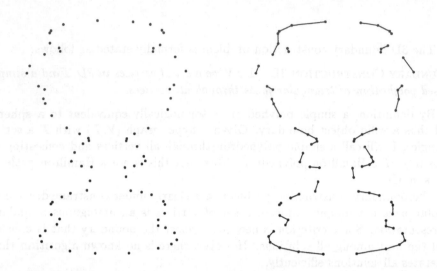

Figure 4.1. Nearest neighbors need not be consecutive along the boundary.

number of triangles in the Delaunay Triangulation is $\Theta(N_v^2)$, giving a worst case complexity of $\Theta(\binom{N_v^2}{2N_v-4})$ which is $\Omega(N_v^{3N_v})$.

All the brute force approaches described require exponential time. In order to achieve a polynomial time complexity some heuristic is used that gives a single boundary. The heuristic must be chosen so as to yield a boundary that is considered a likely boundary for the given set of vertices, among all possible solutions. Such heuristic approaches and qualitative descriptions of the visual environment receive growing interest in the computer vision community [QuaVis, 90].

4.3   Overview of boundary construction methods

In this section I will give an overview of existing methods to find a solution to the boundary construction problems, as stated on pages 38 and 39.

4.3.1   Triangulation growth

Several methods expand some initial triangulation until all vertices are included in the boundary polyhedron. [Boissonnat, 82] assumes that the object surface normal and the Gaussian curvature at each vertex is known. The set of vertices is partitioned into subsets of vertices having the same sign of Gaussian curvature. Each subset is triangulated separately using the mapping of a vertex v_i onto the so-called Gauss sphere. The image of the vertex is the point v_i' on the unit sphere pointed to by the unit surface normal after translation to the origin. An intermediate triangulation (initially a single edge) is grown by considering an edge $v_i v_j$ of the boundary polygon of the triangulation. The vertices in the vicinity of the edge and not yet in the triangulation are mapped onto the Gaussian sphere, and the Convex Hull of the image points is constructed. The

Convex Hull triangle $v'_i v'_j v'_k$ defines a new triangle $v_i v_j v_k$.

[Boissonnat, 84a] does not partition the vertices into subsets of equal sign of the Gaussian curvature. The vertices in the vicinity of the edge are now projected onto a tangent plane. The vertex that sees the edge under the largest angle is taken to create a new triangle. Both algorithms have a time complexity of $\Theta(N_v \log N_v)$.

[Choi et al., 88] assume that there is a viewpoint from which all vertices on the original object surface are visible. Surfaces not satisfying the assumption must be split (apparently manually) into parts that do satisfy the condition, so that the original object surface must be known in the first place. The subsets of vertices are projected onto a sphere and triangulated such that the minimum angle of all triangles is maximized, which results effectively in a Delaunay Triangulation on the sphere. The time complexity is $\mathcal{O}(N_v^2)$.

A drawback of all three methods is that additional data (surface normal and Gaussian curvature), or additional information (a viewpoint from where all vertices on the surface are visible), must be known.

4.3.2   Minimal area polyhedron

[O'Rourke, 81] proposes the polyhedron of minimal surface area as the most natural polyhedral boundary of a set of vertices. An argument for this is that many physical surfaces strive for a situation of minimal tension. Since the tension over the surface is proportional to its area, the object will adopt a shape with minimal surface area.

A heuristic algorithm is presented by [O'Rourke, 81], that starts with the Convex Hull, which is the minimal area polyhedron if all vertices lie on the Convex Hull. The internal vertices with the smallest value for

$$\sum (\text{area new triangles})/(\text{area nearest triangle})$$

are added to the boundary one at a time, where the 'nearest triangle' is the face in the current polyhedron that is closest to that internal vertex, and the 'new triangles' are the faces that must be created in order to include that vertex into the new polyhedron. Each time a vertex is included, all pairs of adjacent triangles $v_i v_j v_k$ and $v_j v_k v_l$ will be considered, and flipped into $v_i v_j v_l$ and $v_i v_k v_l$ if that decreases the surface area.

No proper time complexity analysis is given, but assuming a constant number of flips and $\Theta(N_v)$ vertices interior to the Convex Hull, the average case time complexity would give $\mathcal{O}(N_v^2)$.

This heuristic algorithm does not guarantee a surface within a fixed error fraction of the minimal area. Moreover, as shown by [Boissonnat, 84a] the minimal area polyhedron can be an unnatural boundary.

4.3.3   Minimal area change constriction

[Boissonnat, 84b] takes the Delaunay Triangulation and starts with the boundary of it, i.e. the Convex Hull. In 2D, edges are repeatedly deleted from the

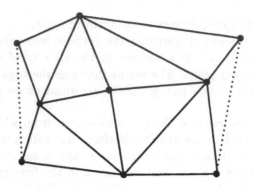

Figure 4.2. Example where constriction of the Delaunay Triangulation will get locked.

intermediate boundary, under the condition that the boundary remains a simple polygon. When a boundary edge $v_i v_j$ is deleted, the adjacent triangle $v_i v_j v_k$ is removed with it, and the inner vertex v_k becomes a part of the new boundary, see Figure 4.2. This is repeated until all vertices are incorporated in the boundary. A similar situation holds in 3D if one triangle of a tetrahedron lies on the intermediate boundary. But also two triangles of a tetrahedron can lie on the boundary. When both triangles are deleted, the adjacent tetrahedron is removed with it, but no new vertices are included in the boundary, since all four tetrahedron vertices were already on the boundary.

In an attempt to minimize the modification of the current boundary, the boundary edge or triangle(s) with the smallest value of

$$(4.1) \qquad \frac{\sum(\text{area interior faces}) - \sum(\text{area boundary faces})}{\sum(\text{area all faces})}$$

is deleted, where in 2D, 'face area' means the length of the edge. This expression yields a small value when the boundary face is small, relative to the inner faces.

Other criteria are possible. For example the maximum distance between faces and the associated part of the circumscribed circle of the boundary triangle or the circumscribed sphere of the boundary tetrahedron [Boissonnat, 84a]. However, this measure is not relative to the size of the triangle or tetrahedron.

The time complexities of this algorithm are $\mathcal{O}(N_v \log N_v)$ in 2D and $\mathcal{O}(N_v^2 \log N_v)$ in 3D.

A drawback of this algorithm is that the Delaunay Triangulation constriction process can get locked, as shown in Figure 4.2. After removing the flattest simplices, no more simplices can be deleted without yielding an invalid boundary. The inner-most vertex can then no longer be included into the boundary, although there exists a closed boundary through all vertices, i.e. a Hamilton polygon, in the Delaunay Triangulation. There also exist Delaunay Triangulations with no Hamilton polygon at all. Both problems are solved in Chapter 5.

4.3.4 Shortest Voronoi Skeleton

Also [O'Rourke et al., 87] construct a simple polygon in the Delaunay Triangulation. Since the geometrically combinatorial dual of any triangulated simple polygon is a tree, the constructed polygon corresponds to a tree in the dual Voronoi Diagram, called Voronoi tree or Voronoi Skeleton. They argue that a natural boundary polygon has a short Voronoi tree, which acts as a skeleton or medial axis. Therefore the minimal length tree in the Voronoi Diagram that corresponds to a dual Hamilton polygon must be found. This idea can also be used in 3D, but in both 2D and 3D the method seems to work properly only for objects with a clear skeleton. Especially in 3D the Voronoi Skeleton is likely to twist a lot in order to reach all Voronoi vertices (or dual tetrahedra), which does not naturally correspond to a skeleton and will give unexpected results. Examples are given in Section 5.6. Note also that the Voronoi vertices, and thus the vertices of the skeleton, need not lie inside the object.

It is not just the shortest tree in the Voronoi Diagram that must be found, but the shortest one that corresponds to a dual Hamilton polygon. The only known algorithm seems to be a trial-and-error algorithm that tries all Voronoi vertices as seed for a tree growth algorithm, and records the shortest tree. This leads to a worst-case time complexity of $\mathcal{O}(N_v^3)$ in 2D, and $\mathcal{O}(N_v^4)$ in 3D.

4.4 Other work

Many other Computational Morphology tasks are related to, but different from, finding a closed boundary through all given points. Some are mentioned in this section.

Clustering
Much work has been done on finding a boundary of a set of points in the plane which generally does not pass through all points. In [Medek, 81], each point has an associated disc touching its nearest neighbor; the union of all discs defines clusters of points. Clustering is also done by the α-Hull (Section 2.2.10).

[Ahuja, 82] performs clustering by grouping Voronoi cells (Section 2.2.9) having similar geometrical properties.

Contour reconstruction from rays
A unique planar contour can be reconstructed from rays [Alevizos et al., 87]. Rays are semi-infinite curves originating at vertices on the contour, and are supposed not to intersect the object. They represent for example the direction from which the points are seen, or the path of a robot arm that has sensed the point. The unique solution can be found in $\mathcal{O}(N_v \log N_v)$ time.

Contour pile
As mentioned in the introduction of this chapter, a boundary can be constructed from a pile of (often parallel) contours by joining vertices on adjacent contours so

as to form triangles. [Keppel, 75] generates the polyhedron of maximal volume. [Fuchs et al., 77] construct the polyhedron of minimal surface area. In contrast to [O'Rourke, 81], their algorithm is non-heuristic due to the knowledge that the vertices lie on contours. A volumetric approach based on the Delaunay Triangulation is presented by [Boissonnat, 88]. There is still research going on to make methods more general, for example in order to handle a different number of contours in adjacent section planes [Ekoule et al., 91].

Mathematical Morphology

The Computational Morphology task that we are interested in here should not be confused with Mathematical Morphology. In Mathematical Morphology objects are considered a collection of subsets of the embedding space (mostly $\mathbb{R}^2$ or $\mathbb{Z}^2$). A lattice and a Boolean algebra are defined on the set of all subsets of the embedding space. Morphological operations are defined in terms of inclusion, union, intersection, and complement, based on logical relations between neighborly set elements, rather than arithmetic ones. Typical operations are shrink operations like erosion and thinning, expand operations like dilation and thickening, and combinations of these like opening and closing [Serra, 86].

It is technically possible to define for example the Delaunay Triangulation constriction operation as a Mathematical Morphology operation [Veltkamp, 89a], but that is rather artificial because the 'vertex objects' are isolated points in the embedding space.

4.5   Concluding remarks

The boundary construction methods mentioned in Section 4.3 can be categorized into volume-based and surface-based methods. The volume-based approaches are the Delaunay Triangulation constriction and the Voronoi Skeleton algorithms, which are based on an *internal* structure on the vertices. The triangulation growth and the minimal polyhedron method are surface-based algorithms, exploiting an external structure on the vertices.

The advantage of volume-based approaches over surface-based ones is that they are potentially more powerful because the internal structure provides more information than an external structure, and often contains the external structure as well.

5

Boundary from the γ-Graph

This chapter presents a boundary construction method based on the γ-Graph. The method is a constriction procedure of the γ-Graph exploiting the inherent geometric information. The weaknesses of other methods are met by the flexibility of the γ-Graph in describing the internal structure of a set of vertices.

5.1 Introduction

The goal of this chapter is to develop a boundary construction algorithm that gets round the weaknesses of the methods described in Section 4.3. As a starting point we take the $\gamma([-1, 1], [0, 1])$ and assume that the set of vertices is non-degenerate, i.e. no $k+2$ vertices lie on an empty ball, so that there are no crossing $(k-1)$-simplices and $\gamma([-1, 1], [0, 1]) = \mathrm{DT}$. This graph describes some internal structure of the set of vertices, is connected and contains not as few joins as the Euclidean Minimum Spanning Tree, but not as many as the complete graph. As a consequence, no additional data is needed for the construction algorithm, as in the triangulation grow methods in Section 4.3.1.

Let the boundary of a 2D graph be a simple polygon of edges in the graph that encloses all vertices and all other edges, and the boundary of a 3D graph a simple closed polyhedron of triangles in the graph that encloses all vertices and all other triangles. An arbitrary graph need not have a boundary, but for example the boundary of any $\gamma([-1, 1], [c_0, 1])$ is the Convex Hull. Constricting the boundary of a graph is the process of deleting a boundary face from the graph, such that a boundary of the new graph is properly defined. A γ-Graph from which faces are deleted is not a γ-Graph anymore, but is called a *pruned* γ-Graph.

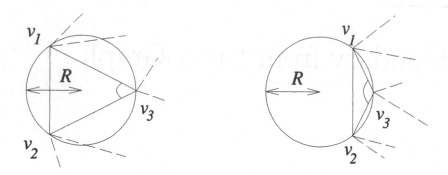

Figure 5.1. Left: γ-indicator > 0. Right: γ-indicator < 0.

In order to find the boundary of a set of vertices, the $\gamma([-1,1],[0,1])$ is constricted on the basis of the geometric information incorporated in the graph, as will be described in Sections 5.2 and 5.3. Situations where the constriction algorithm gets locked or the Delaunay Triangulation contains no Hamilton polygon or polyhedron is dealt with in Section 5.4. The time complexity of the constriction algorithm is analyzed in Section 5.5. In Section 5.6 the results will be compared with two other methods.

5.2   Boundary polygon

This section is concerned with the construction of a boundary polygon of a set of vertices in 2D by constricting the $\gamma([-1,1],[0,1])$ when the set of vertices is non-degenerate, in which case $\gamma([-1,1],[0,1]) = \text{DT}$. Deletion of an edge of a boundary of a (pruned) γ-Graph must keep the boundary a simple polygon. Thus, deletion is only allowed if the vertex opposite to the edge is not already in the current boundary. A boundary edge $v_i v_j$ that satisfies this condition is called *removable* (with respect to v_k). If three edges in the (pruned) $\gamma([-1,1],[0,1])$ implicitly form a triangle, and one of the edges is a boundary edge, then that triangle is called a *boundary triangle*.

The selection of the next removable edge $v_i v_j$ to be deleted is based on the observation that the interior vertex v_k of the boundary triangle $v_i v_j v_k$ that has the largest angle $\angle(v_i v_k v_j)$ has the largest possibility to be seen or sensed from outside the boundary. Additionally, the change of shape of the boundary is small, relative to the size of the triangle, but in another way than defined in Section 4.3.3.

To see how the geometric properties of the $\gamma([-1,1],[0,1])$ can be used in this selection, observe that each edge has two γ-values corresponding with the Delaunay neighborhood. One γ-value of a boundary edge $v_i v_j$ is associated with the empty disc passing through the vertices of the boundary triangle $v_i v_j v_k$. The radius of the disc through v_i, v_j, v_k is denoted by $R(v_i, v_j, v_k)$, and is equal to $r(v_i, v_j)/(1 - |c_0|)$ for some $c_0 \in [-1, 1]$. This γ-value is used in the following definition.

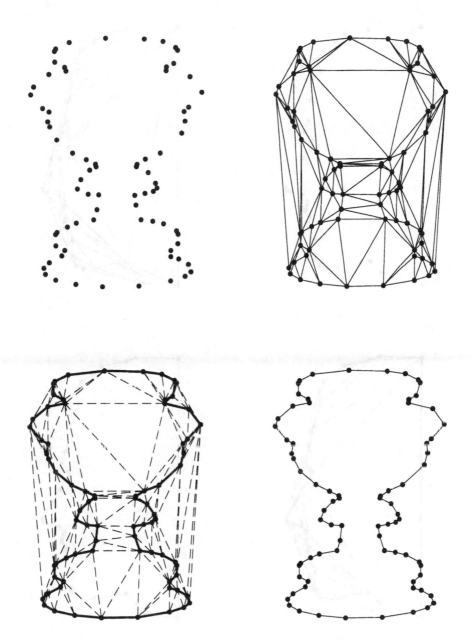

Figure 5.2. Boundary construction example in 2D. Top left: 77 vertices. Top right: the corresponding $\gamma([-1,1],[0,1])$. Bottom left: $\gamma([-1,1],[0,1])$ with the constructed boundary. Bottom right: the constructed boundary.

DEFINITION 5.1 (γ-INDICATOR) *Let $v_i v_j$ be an edge of an intermediate boundary, let $v_i v_k$ and $v_j v_k$ be edges in the (pruned) γ-Graph, and let c_0 be defined by $R(v_i, v_j, v_k) = r(v_i, v_j)/(1 - |c_0|)$. The γ-indicator of $v_i v_j$ with respect to v_k is $|c_0|$ if the center of the circle through v_i, v_j, v_k lies at the same side of $v_i v_j$ as v_k; is $-|c_0|$ if the center lies at the other side; and is zero if $c_0 = 0$.*

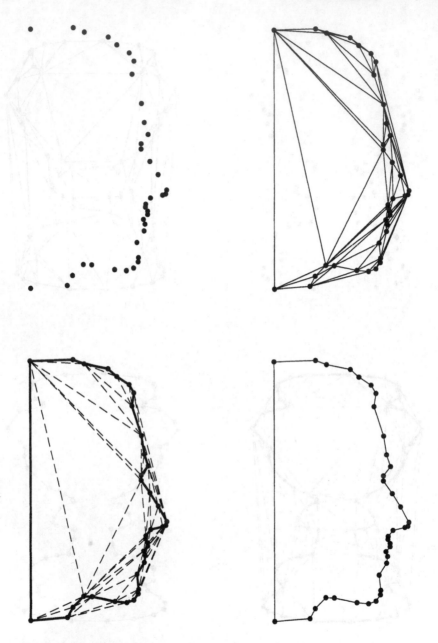

Figure 5.3. Boundary construction example in 2D. Top left: 37 vertices. Top right: the corresponding $\gamma([-1, 1], [0, 1])$. Bottom left: $\gamma([-1, 1], [0, 1])$ with the constructed boundary. Bottom right: the constructed boundary.

The magnitude of the γ-indicator is calculated during construction of the γ-Graph, and can be stored in it. The sign of the γ-indicator is positive if v_i, v_j, and v_k have the same orientation as v_i, v_j and the circle center C, that is, if

$$(5.1) \qquad sign([v_i - v_k, v_j - v_k]) = sign([v_i - C, v_j - C]),$$

where '[ ]' denotes the determinant.

The more negative the γ-indicator, the closer v_k lies to $v_i v_j$, and the larger is angle $\angle(v_i, v_k, v_j)$, see Figure 5.1. Note that the γ-indicator is independent of the size of the triangle. The selection rule based on the γ-indicator is the following:

SELECTION RULE   *Delete the removable boundary edge that has the smallest γ-indicator.*

This selection criterion combines a local measure (the γ-indicator) and global information (the smallest value), and is orientation and scale independent.

Let us now investigate the exact relation between the γ-indicator c_0^i and the angle $\angle(v_i, v_k, v_j)$. According to the sine rule, $\sin(\angle(v_i, v_k, v_j)) = r(v_i, v_j)/R(v_i, v_j, v_k)$. By definition, $r(v_i, v_j)/R(v_i, v_j, v_k) = 1 - |c_0^i|$. If $c_0^i \geq 0$, then $1 - c_0^i = r(v_i, v_j)/R(v_i, v_j, v_k)$, and if $c_0^i \leq 0$, then $1 - c_0^i = 2 - r(v_i, v_j)/R(v_i, v_j, v_k)$. So, if $c_0^i \geq 0$, then $\angle(v_i, v_k, v_j)$ increases when $r(v_i, v_j)/R(v_i, v_j, v_k)$ increases; if $c_0^i \leq 0$, then $\angle(v_i, v_k, v_j)$ increases when $2 - r(v_i, v_j)/R(v_i, v_j, v_k)$ increases. The largest value of $1 - c_0^i$ is obtained for the smallest value of c_0^i, leading to the selection criterion stated above.

Selection and deletion of a boundary edge is repeated until all vertices are part of the boundary polygon. The time complexity of the entire constriction algorithm depends on the way one keeps track of the removable boundary edges and their value of the γ-indicator. Implementational and complexity issues are discussed in Section 5.5.

Figure 5.2 shows the result of the constriction algorithm on a set of vertices of a chalice, and Figure 5.3 shows the result for a set of vertices of a face profile.

5.3   Boundary polyhedron

This section is concerned with the construction of a boundary polyhedron by constricting the $\gamma([-1, 1], [0, 1])$ when the set of vertices in 3D is non-degenerate, so that the triangles in the graph are non-crossing. If four triangles implicitly form a tetrahedron, and one of the triangles is a boundary triangle, the tetrahedron is called a boundary tetrahedron. If $N_v > 4$, a boundary tetrahedron can have one, two, or three boundary triangles. Deletion of triangles from the graph should keep the boundary polyhedron simple. Thus, deletion of three boundary triangles of a boundary tetrahedron is never allowed, deletion of two boundary triangles $v_i v_j v_k$, $v_j v_k v_\ell$ is only allowed if the edge $v_i v_\ell$ is not already in the current boundary (in which case they are called removable with respect to $v_i v_\ell$), and deletion of one boundary triangle $v_i v_j v_k$ is only allowed if the opposite vertex v_ℓ is not already in the current boundary (in which case it is called removable with respect to v_ℓ).

Let us first consider tetrahedra with exactly one boundary triangle. Analogous to the 2D case, the selection of the next removable triangle $v_i v_j v_k$ to be deleted is based on the observation that the opposite vertex v_ℓ of the tetrahedron $v_i v_j v_k v_\ell$ that has the largest solid angle φ has the largest probability to

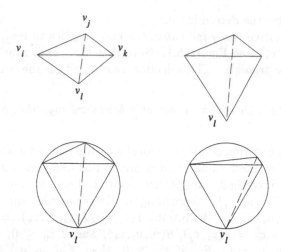

Figure 5.4. The solid angle at v_ℓ in the left column is larger than in the right column. Top row: fixed triangle $v_i v_j v_k$. Bottom row: fixed γ-indicator.

be sensed from outside the boundary. Additionally, the change of shape of the boundary is then small, relative to the size of the tetrahedron, but in another way than defined in Section 4.3.3.

Each triangle in the (pruned) $\gamma([-1,1],[0,1])$ has a Delaunay neighborhood consisting of two empty balls. Let triangle v_i, v_j, v_k have one ball that passes through v_i, v_j, v_k, v_ℓ. The radius of this ball is denoted by $R(v_i, v_j, v_k, v_\ell)$. The selection rule is again based on the notion of γ-indicator:

DEFINITION 5.2 (γ-INDICATOR) *Let $v_i v_j v_k$ be a triangle of an intermediate boundary, let $v_i v_j v_\ell$, $v_j v_k v_\ell$, and $v_i v_k v_\ell$ be triangles in the (pruned) γ-Graph, and let c_0 be defined by $R(v_i, v_j, v_k, v_\ell) = r(v_i, v_j, v_k)/(1 - |c_0|)$. The γ-indicator of $v_i v_j v_k$ with respect to v_ℓ is $|c_0|$ if the center of the sphere through v_i, v_j, v_k, v_ℓ lies at the same side of $v_i v_j v_k$ as v_ℓ; is $-|c_0|$ if the center lies at the other side; and is zero if $c_0 = 0$.*

Like in 2D, the magnitude of the γ-indicator is calculated during construction of the γ-Graph, and can be stored in it. The sign of the γ-indicator is positive if v_i, v_j, v_k, and v_ℓ have the same orientation as v_i, v_j, v_k, and the sphere center C, that is, if

$$(5.2) \qquad sign([v_i - v_\ell, v_j - v_\ell, v_k - v_\ell]) = sign([v_i - C, v_j - C, v_k - C]),$$

where again '[ ]' denotes the determinant.

There is no '3D sine rule' relating the solid angle φ at v_ℓ to $R(v_i, v_j, v_k, v_\ell)$ and $r(v_i, v_j, v_k)$. However, φ does depend on how close v_ℓ lies to $v_i v_j v_k$ relative to the size of the tetrahedron, and on the shape of $v_i v_j v_k$. Observe that $r(v_i, v_j, v_k)/R(v_i, v_j, v_k, v_\ell)$ is independent of the size of the tetrahedron. As in

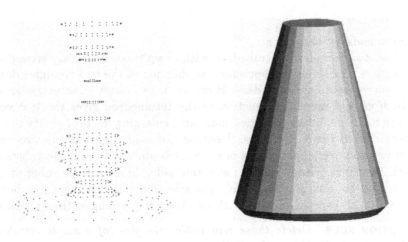

Figure 5.5. Synthetic candlestick object. Top left: 481 vertices in 3D. Top right: Convex Hull, consisting of 73 vertices and 142 triangles. Bottom row: both boundaries consist of 481 vertices and 958 triangles.

2D, if the γ-indicator $c_0^i \leq 0$, then $r(v_i, v_j, v_k)/R(v_i, v_j, v_k, v_\ell) = 1 - c_0^i$, and if $c_0^i \geq 0$, then $2 - r(v_i, v_j, v_k)/R(v_i, v_j, v_k, v_\ell) = 1 - c_0^i$, provided that $v_i v_j v_k$ is fixed. So, the larger $1 - c_0^i$, the larger φ, i.e. the wider the solid angle at v_ℓ. On the other hand, if $1 - c_0^i$ is fixed and the shape of $v_i v_j v_k$ varies, then φ increases when the area A of $v_i v_j v_k$ increases, see Figure 5.4. Since A/R^2 is independent of the size of the tetrahedron, it seems obvious to use a selection criterion based on both $1 - c_0^i$ and A/R^2. However, it appears that using $1 - c_0^i$ alone gives better results. Indeed, a typical Convex Hull contains many triangles of small area (see for example Figures 5.5 and 5.6), and they should be deleted to obtain a good boundary polyhedron.

Let us now consider a tetrahedron with exactly two boundary triangles. All four vertices now lie on the boundary, so deletion of the two triangles does not add a new vertex to the boundary. However, it can result in an extra boundary tetrahedron, and moreover, deletion of the tetrahedron gives the two vertices opposite to the boundary triangles 'more air', enlarging the probability that they are sensed from these directions. Because the solid angles at the two vertices bound two non-overlapping parts of space, it is obvious to sum the values $1 - c_0^i$ of both boundary triangles in the selection rule. Since a large value of $1 - c_0^i$ is equivalent to a small value of c_0^i, the selection rule that captures both the tetrahedra with exactly one and with two boundary triangles then becomes:

SELECTION RULE   *Delete those removable triangles (of a single tetrahedron) that have the smallest sum of γ-indicators.*

Like the selection rule in 2D, this selection criterion combines a local measure (the γ-indicator) and global information (the smallest value), and is orientation and scale independent.

Results of the constriction algorithm are shown in Figures 5.5 and 5.6. Figure 5.5 shows how a synthetic set of 3D vertices modeling a candlestick is processed. Facets that look rectangular are actually two triangles. The bottom left picture shows the result of the constriction algorithm when stopped as soon as all points lie on the boundary. In this case, there are still removable triangles. Especially for such artificial objects there is no general rule telling how long to continue the same constriction procedure when already all vertices lie on the boundary. That decision is typically made interactively by the user. The right picture shows the resulting object after removing just enough triangles. Figure 5.6 illustrates the constriction process performed on points from a laser-range data set of the surface of a mask, measured by [Rioux and Cournoyer, 88]. In this example the algorithm was stopped as soon as all vertices were incorporated in the boundary.

5.4   Hamiltonicity

As shown in Section 4.3.3, the constriction process may stop without finding a Hamilton polygon or polyhedron. Moreover, not every graph contains one. Results from graph theory on Hamiltonicity apply to Hamilton cycles, not Hamilton polyhedra. For example, graph theory surveys the conditions a graph should satisfy to contain a Hamilton cycle, or how efficiently a Hamilton cycle can be found. A very general result is that a graph contains a Hamilton cycle if every vertex has at least $N_v/2$ neighbors [Dirac, 72]. More strict conditions often apply to planar graphs. For example, every four-connected (and thus five-connected) planar graph is Hamiltonian [Tutte, 77]. A special planar graph is a planar triangulation, which is at least two-connected and at most five-connected. A particular planar triangulation is the 2D Delaunay Triangulation, so that we conclude that every four- and five-connected 2D Delaunay Triangulation is Hamiltonian.

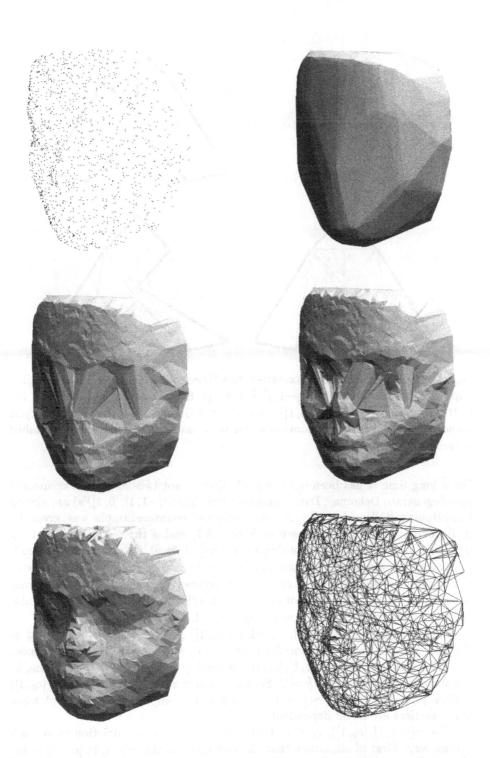

Figure 5.6. Mask reconstructed from laser-range data points. Top left: 1468 scattered vertices. Top right: Convex Hull consisting of 255 vertices and 504 triangles. Middle left: intermediate boundary, 1019 vertices and 2034 triangles. Middle right: intermediate boundary, 1337 vertices and 2670 triangles. Bottom left: final bound-

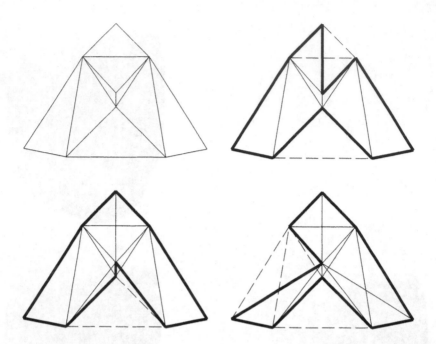

Figure 5.7. Top left: a two-connected non-Hamiltonian Delaunay Triangulation. Top right: the corresponding $\gamma([-1,1],[-0.1,1])$. Bottom left: $\gamma([-1,1],[-0.2,1])$. Bottom right: $\gamma([-1,1],[-0.3,1])$. The three γ-Graphs show the Hamilton cycle found by our constriction algorithm in fat lines, and the deleted edges in dashed lines.

For a long time it has been unknown whether or not two- and three-connected non-degenerate Delaunay Triangulations (and thus $\gamma([-1,1],[0,1])$'s) are always Hamiltonian [O'Rourke, 86]. A two-connected counterexample was given by [Dillencourt, 87], which is shown in Figure 5.7, and a three-connected one by [Dillencourt, 89]. So, a 2D non-degenerate $\gamma([-1,1],[0,1])$ need not be Hamiltonian, and in any case, the constriction process can be unsuccessful. It seems to be unknown whether there exist 3D non-degenerate Delaunay Triangulations, or $\gamma([-1,1],[0,1])$'s, that do not contain a Hamilton polyhedron. However, also in 3D the constriction procedure can get locked.

When varying $c_0 \in [-1,0]$, $\gamma([-1,1],[c_0,1])$ gives a whole spectrum of γ-Graphs. For a c_0 sufficiently smaller than $c_1 < 0$, $\gamma([-1,1],[c_0,1])$ contains more $(k-1)$-simplices than $\gamma([-1,1],[c_1,1])$, implicitly forming more overlapping k-simplices, see Figure 5.7 for $k = 2$. So, for a c_0 sufficiently small, $\gamma([-1,1],[c_0,1])$ is Hamiltonian. After all, $\gamma([-1,1],[-1,1])$ is the complete graph (except when $k + 1$ vertices are linear dependent).

The $\gamma([-1,1],[c_0,1])$, $c_0 \in [-1,0)$, can be used for constriction in the following way. First of all, notice that the boundary of the $\gamma([-1,1],[c_0,1])$ is the Convex Hull. Now consider a 2D $\gamma([-1,1],[c_0,1])$ and a boundary edge v_1v_2 as in Figure 5.8, with boundary triangles $v_1v_2v_3$, $v_1v_2v_4$, and $v_1v_2v_5$. In this

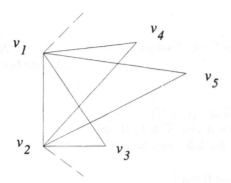

Figure 5.8. Boundary edge in a 2D $\gamma([-1,1],[c_0,1])$, $c_0 \in [-1,0)$.

example, the γ-indicator with respect to v_3 is smaller than those with respect to v_4 and v_5. If $v_1 v_2$ is selected for deletion because it has the smallest γ-indicator of all removable boundary edges, the edges $v_2 v_4$ and $v_2 v_5$ must also be deleted in order to let the pruned γ-Graph have a properly defined boundary. If $v_1 v_2$ is not removable with respect to v_3 but is removable with respect to v_4, and $v_1 v_2$ is selected for deletion due to the γ-indicator with respect to v_4, then $v_1 v_3$ and $v_2 v_5$ must be deleted. In general, if boundary edge $v_i v_j$ is deleted due to the γ-indicator with respect to v_k, any edge crossing $v_i v_j v_k$ must also be deleted.

A 3D $\gamma([-1,1],[c_0,1])$, $c_0 \in [-1,0)$, is constricted in the analogous way. For a c_0 small enough, the γ-Graph will contain more triangles than $\gamma([-1,1],[0,1])$, which implicitly form overlapping tetrahedra. If a boundary triangle $v_i v_j v_k$ is removed due to the γ-indicator with respect to a vertex v_ℓ, any triangle crossing $v_i v_j v_k v_\ell$ must also be removed. We see that any $\gamma([-1,1],[c_0,1])$ can be used for constriction. For a c_0 small enough, the graph will be Hamiltonian, and the process will not get locked.

There is no way of telling in advance for which value of c_0 the graph $\gamma([-1,1],[c_0,1])$ will be Hamiltonian, or constriction will not get stuck. It is possible though to start constriction of $\gamma([-1,1],[0,1])$ and adaptively add $(k-1)$-simplices of a $\gamma([-1,1],[c_0,1])$, $-1 \leq c_0 < 0$, when necessary. This adaptive augmentation will generally be more efficient than constricting $\gamma([-1,1],[c_0,1])$ from the start. On the other hand, the smallest γ-indicator of boundary faces in a $\gamma([-1,1],[c_1,1])$ can be smaller than in a $\gamma([-1,1],[c_0,1])$ for $-1 \leq c_1 < c_0 < 0$, providing better choices for deletion. This is illustrated in Figure 5.7, where some of the extra boundary triangles have larger angles at the interior vertex. Indeed, constriction of $\gamma([-1,1],[-0.3,1])$ gives a different Hamilton polygon than $\gamma([-1,1],[-0.2,1])$ and $\gamma([-1,1],[-0.1,1])$, although $\gamma([-1,1],[-0.1,1])$ is already Hamiltonian.

```
Constrict ()
{ graph Graph;                                    // (pruned) γ-Graph
  (k − 1)-simplex Face, NewFace;                  // 2D: edge; 3D: triangle
  int N_bv;                                       // number of boundary vertices
  heap Heap;

1.     Construct-γ-Graph (Graph);
2.     N_bv = InitialBoundary (Graph, Heap);
3.     while (N_bv < N_v && Heap != φ)
       {
4.        Face = Root (Heap);
5.        if (Removable (Face))
          {
6.           N_bv += Delete (Face, Graph);
7.           for (each NewFace on the boundary)
8.              if (Removable (NewFace))    Insert (NewFace, Heap);
          }
       }
9.     ReportBoundary (Graph);
}
```

Algorithm 5.1. Constriction algorithm.

5.5 Implementation and complexity

The complexity of the constriction algorithm depends on the implementation, and in particular on the data structures. In 2D the γ-Graph is edge-based, so the edges are stored explicitly. The edges incident to a vertex are ordered around that vertex. In 3D the γ-Graph is triangle-oriented and so the triangles are stored explicitly. Edges are also stored explicitly, and the triangles incident to an edge are ordered around that edge.

With these data structures, both triangles and edges can be addressed in constant time, in particular to check whether they lie on an intermediate boundary. Given a boundary face in a pruned $\gamma([-1, 1], [0, 1])$, the γ-indicator can be computed in constant time. Given a boundary face in a pruned $\gamma([-1, 1], [c_0, 1])$, $c_0 < 0$, the smallest γ-indicator can be calculated in $\mathcal{O}(m)$ time, where m is the number of faces incident to the boundary face, which is $\mathcal{O}(N_v)$.

In order to keep track of the boundary faces and their γ-indicators, they are stored in a heap structure sorted on increasing γ-indicator value. In 2D, the root of the heap contains the boundary edge that has the smallest γ-indicator, in 3D it contains the boundary triangles of a single tetrahedron that have the smallest sum of γ-indicators. Fetching the boundary face and revalidating the heap takes $\mathcal{O}(n \log n)$ time for a heap of n elements. In order to keep the heap of size $\mathcal{O}(N_v)$ in the case of a $\gamma([-1, 1], [c_0, 1])$, $c_0 < 0$, only the smallest γ-indicator or sum of γ-indicators of each boundary face is stored, not the values of the overlapping simplices.

Having mentioned the basics, Algorithm 5.1 shows the constriction algorithm in pseudo C-language code. The heap is initially filled with the removable faces on the Convex Hull and their γ-indicator value (line 2); the boundary vertices, edges, and (in 3D) triangles are marked to lie on the boundary, to facilitate the test whether a face is removable. As long as not all the vertices are on the boundary and the heap is not empty (line 3), the face in the root of the heap is taken (line 4), involving revalidating the heap. Although each face is removable at the time it is inserted into the heap, the check in line 5 is necessary since a face can have become unremovable due to deletion of other faces. If deletion is allowed, then the face is deleted (line 6), involving the deletion of overlapping faces in the case of a $\gamma([-1,1],[c_0,1])$, $c_0 < 0$, in order to get an unambiguous boundary. If necessary, the new boundary vertex, edges, and triangles (in 3D), are marked and N_{bv} is incremented when appropriate. Each new boundary face (line 7) is inserted into the heap with its γ-indicator value, if its removal is allowed (line 8). The final boundary can be extracted from the graph (line 9), if desired.

Let us analyze the time complexity of the algorithm for five different cases: the worst case in 2D and 3D for both $\gamma([-1,1],[0,1])$ and $\gamma([-1,1],[c_0,1])$, $c_0 < 0$, and the expected case. Note that the while-loop is executed $\mathcal{O}(N_v)$ times in 2D to include all vertices into the boundary, but $\mathcal{O}(N^2)$ times in 3D, because boundary faces can be removed without adding vertices to the boundary. In the expected case, only the $\gamma([-1,1],[0,1])$ is necessary. The results are listed below, where for line 3 and 7 the number of iterations is given:

| | worst case | | | | exp. case |
| | 2D | | 3D | | 2D and 3D |
line	$\gamma([-1,1][0,1])$	$\gamma([-1,1][<0,1])$	$\gamma([-1,1][0,1])$	$\gamma([-1,1][<0,1])$	$\gamma([-1,1][0,1])$
1	$\Theta(N_v \log N_v)$	$\mathcal{O}(N_v^2)$	$\mathcal{O}(N_v^2)$	$\mathcal{O}(N_v^3)$	$\mathcal{O}(N_v)$
2	$\Theta(N_v \log N_v)$	$\mathcal{O}(N_v^2)$	$\mathcal{O}(N_v \log N_v)$	$\mathcal{O}(N_v^2)$	$\mathcal{O}(N_v \log N_v)$
3	$\mathcal{O}(N_v)\times$	$\mathcal{O}(N_v)\times$	$\mathcal{O}(N_v^2)\times$	$\mathcal{O}(N_v^2)\times$	$\mathcal{O}(N_v)\times$
4	$\mathcal{O}(\log N_v)$	$\mathcal{O}(\log N_v)$	$\mathcal{O}(\log N_v)$	$\mathcal{O}(\log N_v)$	$\mathcal{O}(\log N_v)$
5	$\Theta(1)$	$\Theta(1)$	$\Theta(1)$	$\Theta(1)$	$\Theta(1)$
6	$\Theta(1)$	$\Theta(1)$	$\Theta(1)$	$\Theta(1)$	$\Theta(1)$
7	$\Theta(1)\times$	$\mathcal{O}(N_v)\times$	$\Theta(1)\times$	$\mathcal{O}(N_v)\times$	$\Theta(1)\times$
8	$\mathcal{O}(\log N_v)$	$\mathcal{O}(\log N_v)$	$\mathcal{O}(\log N_v)$	$\mathcal{O}(\log N_v)$	$\mathcal{O}(\log N_v)$
9	$\Theta(N_v)$	$\Theta(N_v)$	$\Theta(N_v)$	$\Theta(N_v)$	$\Theta(N_v)$
tot	$\mathcal{O}(N_v \log N_v)$	$\mathcal{O}(N_v^2 \log N_v)$	$\mathcal{O}(N_v^2 \log N_v)$	$\mathcal{O}(N_v^3 \log N_v)$	$\mathcal{O}(N_v \log N_v)$

All storage complexities are dominated by the size of the γ-Graph, which is given by Lemmas 3.4 and 3.5.

For practical cases, the use of $\gamma([-1,1],[0,1])$ is predominant, resulting in a worst-case time complexity of $\mathcal{O}(N_v \log N_v)$ for 2D and $\mathcal{O}(N_v^2 \log N_v)$ for 3D. Note, however, that the latter complexity stems from the worst possible situation, i.e. the number of triangles in the γ-Graph is $\mathcal{O}(N_v^2)$. In terms of the number of triangles N_t, line 1 takes $\mathcal{O}(N_t \log N_t)$ (see Section 2.2.9), and the iteration over line 3 is performed $\mathcal{O}(N_t)$ times, resulting in a total of $\mathcal{O}(N_t \log N_t)$ worst case time complexity.

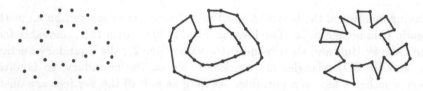

Figure 5.9. Left: set of vertices from [O'Rourke et al., 87]. Middle: shortest Voronoi Skeleton result. Right: constriction result.

Algorithm 5.1 has been implemented in C. The constriction process takes about four seconds on a Sun SparcStation 1+ for the mask data set of Figure 5.6. The $\gamma([-1, 1], [0, 1])$ on that set of vertices consists of 18274 triangles, forming 8633 tetrahedra.

5.6 Comparison

Once we have the γ-Graph constriction algorithm, the minimal area change constriction of the Delaunay Triangulation and the Voronoi Skeleton method (see Section 4.3) are easily implemented, because both methods are based on the Delaunay Triangulation, that is, the $\gamma([-1, 1], [0, 1])$. It turns out that in 2D all three methods often give the same result. A set of vertices that gives different results, taken from [O'Rourke et al., 87], is shown in Figure 5.9. We see that if the original object is very curled, it is not likely that the vertices are sensed from some distance of the object, and the γ-indicator provides no proper

Figure 5.10. Comparing three methods on a set of vertices from a bottle's surface. Left: γ-indicator method. Middle: minimal area change method. Right: Voronoi Skeleton method.

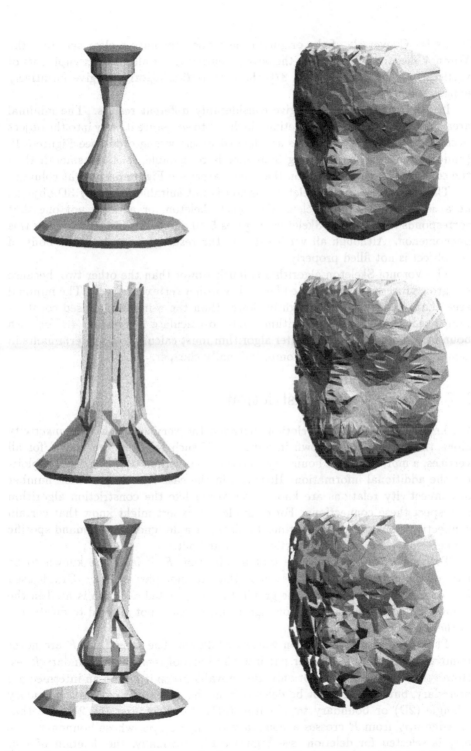

Figure 5.11. Comparing three methods on the candlestick (left column) and mask (right column) data set. Top: γ-indicator method. Middle: minimal area change method. Bottom: Voronoi Skeleton method.

heuristic. Conversely, if the original object does not have a clear skeleton, the Voronoi Skeleton method uses the wrong heuristic. For all other example sets of vertices from [O'Rourke et al., 87], the constriction algorithms give intuitively expected boundaries.

In 3D the three methods give considerably different results. The minimal area change constriction is sometimes inclined to sculpture its way into the object because the boundary triangles are deleted in the wrong order, see Figure 5.10 (middle). Even if the resulting boundary is reasonable, it is less smooth than the constriction result based on the γ-indicator: see Figure 5.11, right column.

The shortest Voronoi Skeleton method is not suitable for many 3D objects, because there often is no clear 3D object skeleton, or at least not one that corresponds to a Voronoi Skeleton. Figures 5.10 and 5.11 show examples of this phenomenon. Although all vertices lie on the resulting boundary, the body of the object is not filled properly.

The Voronoi Skeleton algorithm is much slower than the other two, because the grow procedure is performed for each Voronoi vertex as a seed. The minimal area change constriction is slightly slower than the γ-indicator-based constriction, because the former algorithm needs to calculate Expression 4.1 for each boundary simplex while the latter algorithm must calculate the determinants in Equation 5.1 or 5.2, which is computationally cheaper.

5.7   Constrained constriction

We have assumed that no relations between the vertices, such as connectivity along the boundary, is known in advance. If such relations are known for all vertices, a more powerful boundary construction method could be used, exploiting the additional information. However, in the case that only a small number of connectivity relations are known, we would like the constriction algorithm to respect these connections. For example, an expert might know that certain connections must exist in experimental data, or a designer may demand specific connections when specifying vertices of an artifact.

To be more precise, assume that a collection F of faces are known to be part of the boundary. The faces of F that are not part of the γ-Graph used for constriction are added to the graph; this augmented γ-Graph is used in the following constrained constriction algorithm, which is not allowed to delete any of the faces from F.

The constrained constriction works as follows. The faces from F are never removable and therefore never put into the heap of removable boundary faces. However, a face from F is not only unremovable when it lies on an intermediate boundary, but may also not be deleted from the graph if it crosses a boundary triangle (2D) or boundary tetrahedron (3D). Consider first the 2D case that an edge $v_j v_\ell$ from F crosses a boundary triangle $v_i v_j v_k$ whose boundary edge $v_i v_j$ is selected for deletion, see Figure 5.12. Normally, the deletion of $v_i v_j$ involves the deletion of $v_j v_\ell$ from the graph, but $v_j v_\ell$ must remain in the graph. Instead, $v_i v_j$ and the γ-indicator with respect to v_k is deleted from the heap,

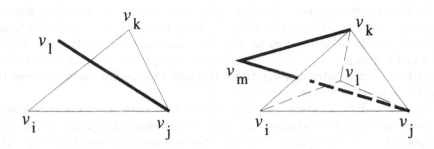

Figure 5.12. Constrained constriction: the bold edge (left) and the bold triangle (right) may not be deleted.

and to prevent reinsertion, $v_i v_k$ is deleted from the graph. If $v_i v_\ell$ is not already present, it is inserted into the graph, so that $v_i v_j v_\ell$ is a boundary triangle. If $v_i v_j$ is removable with respect to v_ℓ, it is inserted into the heap together with the γ-indicator with respect to v_ℓ. In one of the next iterations, $v_i v_j$ may be deleted from the graph and the heap, and bring $v_j v_\ell$ into the boundary. However, a constrained constriction may get locked if $v_j v_\ell$ is not yet in the boundary and $v_i v_j$ has become unremovable. Note that apart from $v_j v_\ell$, other edges may cross $v_i v_j v_k$ as well; however, that does not effect the constrained constriction presented above.

The analogous 3D situation is depicted at the right in Figure 5.12, where $v_i v_j v_k$ is a boundary triangle and $v_j v_k v_m$ a triangle from F. Normally, the deletion of $v_i v_j v_k$ involves the deletion of $v_j v_k v_m$, but $v_j v_k v_m$ must remain in the graph. Instead, $v_i v_j v_k$ and the γ-indicator with respect to v_ℓ is deleted from the heap, and to prevent reinsertion into the heap, $v_i v_j v_k$ is also deleted from the graph. If $v_i v_j v_m$ and $v_i v_k v_m$ are not already present, they are inserted into the graph, so that $v_i v_j v_k v_m$ is a boundary tetrahedron. If $v_i v_j v_k$ is now removable with respect to v_m, it is inserted into the heap together with the γ-indicator with respect to v_m. Like in 2D, a constrained constriction may get locked when $v_j v_k v_m$ is not yet in the boundary but has become unremovable. Apart from $v_j v_k v_m$ other triangles may cross $v_i v_j v_k v_\ell$ as well, but that does not effect the constrained constriction.

The triangles at the back of the mask in Figure 5.6 where prevented from deletion by constrained constriction.

5.8 Concluding remarks

In this chapter, the geometric information contained in the γ-Graph is used to construct a closed piecewise linear object boundary through scattered points. The γ-Graph on the set of points is successively constricted until the boundary of the pruned γ-Graph is a proper object boundary, passing through all vertices. While constriction of the Delaunay Triangulation may stop without having found a Hamilton polygon or polyhedron, the parameters of the γ-Graph provide the flexibility to find a proper boundary. The selection of the boundary faces to remove is based on the combination of a local measure (the γ-indicator), and global information (the minimum value). This criterion yields good looking boundaries, compared to the minimal area change and the Voronoi Skeleton algorithm. Our constriction algorithm is easily extended so as to prevent the deletion of a priori known boundary faces.

It is not widely acknowledged that construction methods based on a geometric graph (such as the Delaunay Triangulation, Voronoi Diagram, or γ-Graph) are generally not consequent. That is, if the constructed boundary of a vertex set V is B and a new vertex that lies on B is added to V, the new boundary need not be B. This results from the fact that in general the geometric graph of the new vertex set does not contain all the edges or triangles of the graph on V, in particular those of the boundary segments of B.

Another interesting case is the behavior of the algorithm when the vertices are distributed arbitrarily dense over a known object boundary. The object boundary segments then become arbitrarily small. When the limit is reached, the γ-indicator of each segment is one, but the γ-indicator of intermediate boundary segments is smaller than one. So, the object boundary segments are never removed, and therefore the constructed boundary is the known object boundary.

The result of our constriction process is a pruned graph. A resulting pruned $\gamma([-1, 1], [0, 1])$ implicitly defines a triangulation of the interior of the object. A pruned $\gamma([-1, 1], [c_0, 1])$, $c_0 < 0$, can easily be pruned further to obtain a triangulation of the interior. Such a triangulation of the interior can be used to calculate properties such as the volume and mass of the object.

We have seen that an object skeleton need not be a good tool to construct a boundary, but conversely the triangulation or tetrahedralization of the pruned γ-Graph always provides *some* skeleton of the object, see Section 4.3.4.

The vertices are assumed to lie on the boundary of an object without holes, but an inner contour or surface can be handled separately. The triangulation of the interior of the outer boundary does not correspond with the body of the object anymore, but for the calculation of some properties the value corresponding to the inner boundary can be subtracted from the result corresponding to outer boundary. The case of objects with handles is considered difficult, and is a possible subject of further research.

6

Approximation and localization

This chapter introduces the problem of (hierarchical) approximation and localization of polygonal and polyhedral objects, and presents several approximation error criteria. An overview of existing boundary-based intrinsic schemes as well as some other work on approximation and localization are given.

6.1 Introduction

Two facilities are often used for efficient manipulation of complex polygonal or polyhedral objects consisting of many faces: approximation and localization, which can both be performed hierarchically. The purpose of both techniques is to avoid unnecessary processing of much detail.

If the vertices of the polygon or polyhedron lie on a regular grid, finding an approximation or localization is a simpler task than for an arbitrary vertex connectivity structure, or topology. For example [Schmitt and Gholizadeh, 86] provide an efficient approximation scheme for triangular polyhedra whose vertices lie on a regular grid. In the following we will only consider approximation and localization schemes for arbitrary topologies.

Approximation
It is not always necessary to process an object in full detail. To obtain quick approximate results an approximation of the object may often be used. If the approximation object consists of much fewer faces than the original object, the processing is usually much faster.

For example, the interactive manipulation of a 3D object requires real time display. During animated rotation, an approximation object allows faster display without much loss of reality. When a fixed viewpoint is chosen, the object can be displayed in full detail again. Another example is the perspective display of far away objects in 3D that are mapped to only few pixels. The details of the object are then not visible, and an approximation of the object suffices, see also [Clark, 76].

Another reason for using levels of approximation is that features of an object can be classified by following the features through successively more coarse approximations of the object. This way of feature classification is used in object recognition.

Localization

Localization provides information about the position of the object, or the boundary of the object, by means of a set of bounding volumes that together contain (the boundary of) the object. If the bounding volumes allow efficient testing, operations such as point location and intersection tests can be performed efficiently. For example, if two objects are both known to lie in a sphere, the objects cannot intersect if the spheres do not intersect, which can be efficiently tested.

Hierarchy

Both approximation and localization can be performed at several levels of detail. If the successive levels are such that a segment at one level is refined at the next level, there is a hierarchy of successively more detailed levels. A hierarchy is naturally stored in a tree data structure, where the root contains the most coarse level of approximation, and the sons of a node represent the refinement of the parent at the next level of approximation. Algorithms operating on the hierarchy try to solve their task at the root. If that is not possible, the algorithm proceeds at successively more detailed levels.

If approximation and localization are combined, the bounding volumes are associated with an approximation face, and localize the part of the object that is approximated by that face. The advantages of approximation, localization, and a hierarchy can be combined into a single scheme. This is illustrated by the following example. A point-in-polygon or point-in-polyhedron test determines whether a given point X is internal to a given polygon or polyhedron P. One way to decide this is to count the number of intersections between P and any half-line originating from X, see e.g. [Preparata and Shamos, 85]. If X is not on P, it is internal to P if the number of intersections is odd, and external otherwise. In order to count the number of intersections, one has to test the half-line against each boundary segment of P.

This test can be performed more efficiently if we can use an approximation of P that yields the same answer to the test as P itself. The next lemma tells when the approximation of a part of P does not affect the inclusion test.

LEMMA 6.1 *Let B be a part of P, A an approximation of B connected with the rest of P without cracks, and F a bounding volume containing A and B. If X is external to F, then X is internal/external to P if and only if X is internal/external to P with B replaced by A.*

Proof. If X is external to F, then X does not lie between A and B. Therefore, the replacement does not change the result of the inclusion test. □

The hierarchical point location test starts at the root of the tree by testing if X is external to the bounding volume. If so, it is also external to P. Otherwise the algorithm proceeds at the next level. If X lies outside a bounding volume, the approximation segment at that level can be used to perform the location test. Otherwise the algorithm proceeds locally at the next level.

Schemes for approximation and localization can be classified into volume-based and boundary-based methods. The former models represent the object's enclosed volume, the latter ones represent the boundary of the object. Another classification can be made into domain-dependent models and intrinsic models. Domain-dependent methods are based on a decomposition of the embedding space according to a predefined grid. The intrinsic schemes are based on the shape of the object. These two orthogonal classifications give four possible combinations.

An advantage of intrinsic schemes over domain-dependent ones is the independence of the orientation of the object. If an object is affinely transformed, the approximation and the bounding volumes are transformed in the same way in the case of an intrinsic scheme, whereas a domain-dependent representation has to be constructed again from scratch. This thesis concerns boundary construction and manipulation, so therefore we are interested in boundary-based schemes. More specifically, we want a hierarchical approximation and localization scheme that has a uniform definition in 2D and 3D, and bounding areas and volumes that allow efficient testing.

Section 6.2 presents some approximation error criteria, which provide a measure for expressing how good an approximation is. Section 6.3 gives an overview of existing boundary-based intrinsic schemes, and Section 6.4 gives examples of other schemes.

6.2 Error criteria

The error of an approximation can be defined in several ways. For the approximation of an arbitrary arc-length parameterized plane curve $C(t)$ by an arbitrary arc-length parameterized approximation curve $C_a(t)$, a possible measure for the error is the maximum distance between points on the curves at corresponding parameter values: $\max_t d(C(t), C_a(t))$. If $C(t)$ is not an arbitrary curve but a polyline $v_p \ldots v_s$, the maximum error always occurs at one of the vertices. If additionally the approximation curve is a line segment g, the above approximation

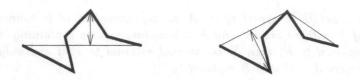

Figure 6.1. The error in a hierarchical approximation may increase.

error becomes

(e1)
$$\max_{p \leq i \leq s} d(v_i, g).$$

A variation is obtained by considering the distance to the line through g instead of g itself:

(e2)
$$\max_{p \leq i \leq s} d(v_i, line(g)).$$

In a hierarchy of approximations the successive levels contain more and more vertices. However, in the case that the vertices of the approximation polyline must be vertices of the original, the error of successive approximations need not decrease uniformly towards zero, and may even increase, see Figure 6.1.

The above error criteria are extended to 3D in a straightforward manner: the line segment g in Error (e1) is replaced by a triangle, and $line(g)$ in Error (e2) is replaced by $plane(g)$, the plane through triangle g.

Since the bounding volumes in a boundary localization scheme form a covering of the boundary, it can also be regarded as an approximation, having an associated error. In general, the definition of the approximation error of a covering depends on the specific bounding volumes. I will confine myself to mentioning two examples that are considered by [Imai and Iri, 88]. The first one is a covering of planar polylines by so-called strips:

DEFINITION 6.1 (STRIP) *A strip that covers $v_p, \ldots, v_s$ is the minimum area rectangle containing $v_p, \ldots, v_s$ with two sides parallel to the line through v_p and v_s.*

Figure 6.2 shows a strip covering a polyline, and two smaller strips covering parts of the polyline. Note that the sides of the strip need not pass through v_p and v_s. A strip has two widths w_1 and w_2 that are the distances from the line through v_p and v_s to the parallel sides of the rectangle. The error associated with a single strip, say strip i, is defined as $(w_{i,1} + w_{i,2})/2$, and the error of approximation of a polyline by m strips is

(e3)
$$\max_{1 \leq i \leq m} (w_{i,1} + w_{i,2})/2.$$

Another covering can simply be made with the minimum area rectangle containing $v_p, \ldots, v_s$. The width w of this rectangle is the length of the smallest side.

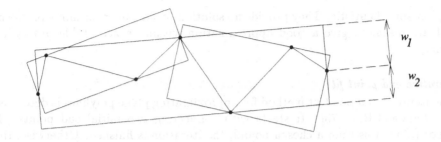

Figure 6.2. Strips covering a polyline.

The associated approximation error of a single rectangle i is $w_i/2$, and the error of approximation of a polyline by m rectangles is

(e4)
$$\max_{1 \leq i \leq m} w_i/2.$$

Note that the Errors (e3) and (e4) are *approximation* errors, not localization errors. A bounding volume just bounds the position of something. Naturally, the smaller the bounding volume, the tighter the localization, but one cannot speak of a localization error.

6.3   Boundary-based intrinsic schemes

Optimal approximation
The min-# approximation problem for a polygonal curve in the plane is the problem of finding an approximation polygon with the minimal number of vertices and with the error within a given bound. The min-ϵ problem is the problem of finding an approximation polygon with the minimal error and a given number of vertices. Algorithms for both optimality problems are discussed by [Imai and Iri, 88]. The time complexities of the best available algorithms are as follows:

error criterion	min-#	min-ϵ
(e1)	$O(N_v^2 \log N_v)$	$O(N_v^2 (\log N_v)^2)$
(e2)	$O(N_v^2 \log N_v)$	$O(N_v^2 \log N_v)$
(e3)	$O(N_v^2 \log N_v)$	$O(N_v^2 \log N_v)$
(e4)	$O(N_v \log N_v)$	$O(M_v N_v (\log N_v)(\log(M_v/N_v)))$

in which M_v is the number of vertices in the approximation polygon.

These complexities apply to a single approximation. A sequence of successively more detailed approximations requires the iterative application of the algorithm. However, in general this yields no hierarchy of approximations, and is computationally expensive.

No efficient algorithms seem to be available for these optimality problems in 3D. In order to avoid excessive time complexities, all the following schemes exploit some heuristic. They provide no solution to the min-# or min-ϵ problem, but are meant to give a good approximation within an acceptable processing time.

Iterative end point fit

The iterative end point fit method for approximating plane polylines is described in [Duda and Hart, 73]. It starts with connecting two initial end points. If Error (e1) is less than a chosen bound, the iteration is finished. Otherwise, the vertex that determines the error is connected with the two end points, yielding two new line segments. The process is repeated for the new line segments. This method is also known, especially in the cartography community, as the *Douglas–Peucker* algorithm, after [Douglas and Peucker, 73].

The time complexity for a hierarchical approximation up to the most detailed level is $\Theta(N_v \log N_v)$, which is optimal [Hershberger and Snoeyink, 92].

Strip tree

The strip tree [Ballard, 81] is a binary tree that stores all intermediate strips associated with the approximation segments resulting from the iterative end point fit method algorithm. The root of the strip tree contains the strip covering all the points. The sons are each root of a subtree covering $v_p, \ldots, v_q$ and $v_q, \ldots, v_s$, where v_q is a vertex selected by the iterative end point fit method. The construction of the strip tree stops when Error (e3) is within a chosen bound. Apparently two different error criteria are used: (e1) to select a new point to construct a strip, and (e3) to decide to stop. Obviously, the time complexity is equal to that of the iterative end point fit algorithm.

A variant of the strip tree is the Binary Line Generalization (BLG) tree [Oosterom and Bos, 89], used for Geographic Information System applications. In the BLG tree the node containing a vertex v_q does not store the corresponding strip, but the distance to line segment $v_p v_s$.

Arc tree

The arc tree [Günther, 88] is a balanced binary tree for representing a hierarchy of approximations of arbitrary plane curves. Let a curve C with length l be parameterized by its arc length fraction $t \in [0, 1]$. Then the length of the curve from $C(0)$ to $C(t_0)$ is $t_0 l$. The j-th approximation consists of 2^j line segments. Line segment i connects the points $C(i/2^j)$ and $C((i+1)/2^j)$, and is the approximation of the arc of C between $C(i/2^j)$ and $C((i+1)/2^j)$, which has length $l/2^j$. The hierarchy of successive approximations is stored in a binary tree.

The sequence of polygonal approximations converges uniformly towards $C(t)$ with respect to Errors (e1) and (e2). The following nice property is advantageous for a hierarchical point location test and for intersection operations. The arc between $C(i/2^j)$ and $C((i+1)/2^j)$ is contained in the ellipse whose focal points are the two end points of the arc. At a higher resolution the number of bounding ellipses increases, but their total area decreases, thus providing a better localization.

Figure 6.3. 3D iterative extreme point fit algorithm. Left: triangle 3-split and vertex for 2-split. Middle: subsequent triangle 2-split. Right: overall hierarchy tree.

Despite its elegance, the arc tree has several drawbacks. Because the parameterization by arc length does not generalize to surfaces in 3D, the arc tree has no higher-dimensional equivalent. Computing intersections of bounding ellipses in order to perform hierarchical intersection operations is expensive. Finally, in this thesis we are concerned with the approximation of a *polygonal* curve C. In that case, the points $C(i/2^j)$ need not be vertices of polygon C. A variant scheme, the *polygon arc tree*, breaks each line segment approximating polyline $v_i \ldots v_{i+n}$ into two ones approximating $v_i \ldots v_{i+\lceil n/2 \rceil}$ and $v_{i+\lceil n/2 \rceil} \ldots v_{i+n}$, respectively. Obviously, this need not be a good approximation. Moreover, Errors (e1) and (e2) need not decrease uniformly anymore.

The time complexity for building the polygon arc tree is $\Theta(N_v \log N_v)$ in the best case, and $\mathcal{O}(N_v^2)$ in the worst case.

Delaunay pyramid

The Delaunay pyramid [DeFloriani, 89] is a representation of a sequence of 2D Delaunay Triangulations. The Delaunay pyramid is used to represent terrain surfaces, where a height attribute is associated with data points (also called $2\frac{1}{2}$D surfaces). All the triangulations are done in the plane. Starting with an initial Delaunay Triangulation, a number of data points is added to obtain a more accurate terrain surface approximation. The corresponding Delaunay Triangulation is the next one in the sequence. Such a sequence of Delaunay Triangulations together with a set of links describing the successive changes in the triangulation constitute the Delaunay pyramid. Since all triangulations are done in the plane, this scheme is unsuited to represent closed boundaries of 3D objects.

The best case and the expected case time complexity to build the Delaunay pyramid is $\Theta(N_v)$. The worst case complexity is $\mathcal{O}(N_v^2)$.

3D iterative extreme point fit

A 3D analogue of the 2D iterative end point fit approximation, which I shall call the 3D iterative extreme point fit, is presented by [Faugeras et al., 84]. The algorithm starts with a single triangle approximating the whole polyhedron. It then finds a closed path of edges in the polyhedron that contains the three vertices of the first triangle. This path divides the polyhedron into two open polyhedra. The algorithm now iterates as follows. Given a triangle and an open

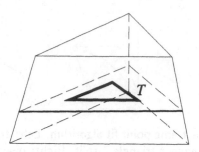

Figure 6.4. Prism.

polyhedron that is approximated by this triangle, the vertex in the polyhedron having the largest distance to the triangle is connected to the three triangle vertices, giving three new triangles.

If the triangles are always split into three new ones, the new triangles get more and more elongated, and the edges remain in all next levels of approximation. Therefore, after each iteration that splits triangles into three triangles an iteration follows that splits each pair of new triangles that have an *old* edge in common at a vertex on the associated path, see Figure 6.3. In that way, each of these triangles is split into two. Elongated triangles thus cannot always be avoided, but they are removed at a next level of approximation.

The approximating triangles are naturally stored in a hierarchy tree with alternating levels of nodes having three and two children.

The time complexity for a hierarchical approximation up to the most detailed level depends on which vertices lie furthest from the approximation triangle. The best case complexity is $\mathcal{O}(N_v(\log N_v)^2)$, the worst case complexity is $\mathcal{O}(N_v^2 \log N_v)$.

Prism tree

The prism tree [Ponce and Faugeras, 87] is a localization scheme that stores bounding volumes of the open polyhedra approximated by the 3D iterative extreme point fit algorithm. The bounding volumes are truncated pyramids which are called prisms by [Ponce and Faugeras, 87]:

DEFINITION 6.2 (PRISM) *Let T be an approximation triangle and P the approximated open polyhedron. The prism is a truncated pyramid consisting of five faces: a top, a bottom, and three sides. The top and the bottom are triangles parallel to T. The sides are quadrilaterals parallel to the three bisector planes of T and its three neighboring approximation triangles. With orientation of the faces thus determined, the prism is the smallest truncated pyramid containing P.*

Analogous to the sides of the strips, the prism sides parallel to the bisector planes need not pass through the edges of the triangle, see Figure 6.4.

Not T itself, but the truncated pyramid bounding P is stored in a node of the tree. A level in the tree corresponds to a collection of truncated pyramids enclosing the boundary, rather than to an approximation polyhedron.

Obviously, the time complexity is equal to that of the 3D extreme point fit algorithm.

6.4   Other schemes

For comparison, this section gives a few examples of schemes that are not boundary-based or intrinsic.

Delaunay tree

The kD Delaunay tree [Boissonnat and Tellaud, 86] stores all temporary k-simplices resulting from the incremental construction of the Delaunay Triangulation of the vertices of the object. The resulting tree is a hierarchical volumetric representation that allows fast point location. The Delaunay tree itself is not an object representation. In order to use it for that purpose, all the object faces should belong to the Delaunay Triangulation of the set of vertices. Additionally, the simplices inside and outside the object need to be determined. To use the Delaunay tree for representing approximations is even harder. In that case, the order in which the points are added must correspond to successive approximations.

Sphere decomposition

The sphere decomposition method by [O'Rourke and Badler, 79] is volume-based, intrinsic, and non-hierarchical. The method selects boundary vertices, and then fits spheres passing through these vertices. The union of the spheres need not completely cover the whole object. Therefore, these spheres cannot be used as bounding volumes. Moreover, the representation is not hierarchical. In order to yield a good approximation of the shape of the object, many overlapping spheres are needed.

Quadtree and octree

All previous schemes are intrinsic. The quadtree [Samet, 84] and the octree [Meagher, 82] are domain-dependent representations. They are hierarchical and volume-based.

The quadtree starts with an initial rectangle containing a plane polygon. This rectangle is symmetrically split into four equally sized subrectangles, and each subrectangle is recursively subdivided if it is not completely inside or outside the polygon. The recursion stops when the rectangles have some predefined minimum size. All rectangles are labeled with 'empty', 'full', or 'partly filled'.

Of course the quadtree is not limited to polygonal objects, but can be used for arbitrarily defined objects. The union of the full rectangles as well as the union of both the full and partly filled rectangles form an approximation of the interior of the object. The union of both the full and partly filled rectangles also forms a bounding volume.

The octree is analogously defined for 3D objects, subdividing parallelepipeds into eight equally sized subparallelepipeds. Since both representations subdivide according to a predefined grid, they do not take into account the shape of the object. For arbitrary polygonal or polyhedral objects, only extensive subdivisions give good approximations.

6.5   Concluding remarks

In this thesis we are interested in boundary construction and manipulation, and therefore in boundary-based approximation and localization schemes. Furthermore, intrinsic schemes, unlike domain-dependent ones, are independent of the orientation of the object. The boundary-based intrinsic schemes mentioned in this chapter are summarized in the following table, showing whether the scheme is 2D or 3D ($2\frac{1}{2}$D for the Delaunay pyramid), and whether the scheme is an approximation scheme (A), a localization scheme (L), and/or hierarchical (H):

scheme	reference	2D	3D	A	L	H
optimal appr.	[Imai and Iri, 88]	+		+		
2D fit	[Duda and Hart, 73]	+		+		+
strip tree	[Ballard, 81]	+			+	+
arc tree	[Günther, 88]	+		+	+	+
Delaunay pyr.	[DeFloriani, 89]		+	+		+
3D fit	[Faugeras et al., 84]		+	+		+
prism tree	[Ponce and Faugeras, 87]		+		+	+
flintstones	Chapter 7	+	+	+	+	+

The flintstone scheme, to be presented in the next chapter, is also included for comparison. As indicated in the table, it is a scheme for hierarchical approximation and localization in both 2D and 3D. More specifically, it has a uniform definition in 2D and 3D, and bounding areas and volumes that allow efficient testing operations.

Approximation of digitized curves is often used to remove digitization errors and smooth the curve [Ray and Ray, 92]. Here, we assume that all vertices are correct and lie on a closed object boundary. At the level of full detail all vertices must be present and the topology must be the same as the original one.

7

The flintstones

This chapter presents a new scheme for hierarchical approximation and localization of closed polygonal and polyhedral objects without holes. The bounding volumes, called flintstones, are defined by discs in 2D and balls in 3D. This makes the flintstones storage efficient, and operations on flintstones computationally cheap.

7.1 Introduction

The goal of this chapter is to develop a hierarchical approximation and localization scheme whose definition is naturally generalized from 2D to 3D, and is computationally efficient for hierarchical operations such as intersection and point-in-object tests. This chapter introduces the flintstone scheme, so called because of the shape of the bounding areas. The approximation algorithms are based on the definitions of the bounding area and volumes which are composed of discs or balls, respectively. These algorithms and their time complexities are presented in Sections 7.2 and 7.3. Section 7.4 presents how the flintstone scheme can efficiently be used for hierarchical operations, in particular point-inclusion and intersection operations.

7.2 Flintstones in 2D

In this section we consider the approximation and localization of a closed 2D polygon $v_0 \ldots v_{N_v-1}$. The approximation algorithm is based on the way the localization of a part of the closed polygon, an open polyline $v_p \ldots v_s$, is performed.

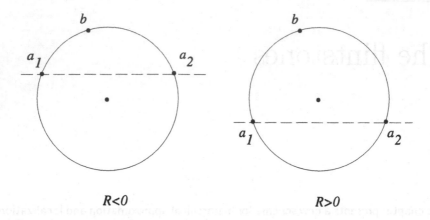

Figure 7.1. Disc $D(a_1, a_2; b)$ with signed radius.

7.2.1   Localization

The definition of the bounding area is based on the notion of a disc with a signed radius:

DEFINITION 7.1 (DISC WITH SIGNED RADIUS) *Let a_1, a_2 and b be points not lying on one line. The radius of the disc $D(a_1, a_2; b)$ touching these points is negative if the center of the disc and b lie at opposite sides of the line through a_1 and a_2, and positive otherwise.*

This definition is illustrated in Figure 7.1.

DEFINITION 7.2 ($\pm$-OPERATOR) *Let B_1 and B_2 be two balls with signed radii R_{B_1} and R_{B_2}, respectively. The $\pm$-operator is defined as*

$$B_1 \pm B_2 = \begin{cases} B_1 \cap B_2 & \text{if } R_{B_1}, R_{B_2} < 0, \text{ or} \\ & \quad R_{B_i} < 0, R_{B_j} > 0, |R_{B_i}| > R_{B_j}, i, j = 1, 2, i \neq j, \\ B_1 \cup B_2 & \text{if } R_{B_1}, R_{B_2} > 0, \text{ or} \\ & \quad R_{B_i} < 0, R_{B_j} > 0, |R_{B_i}| < R_{B_j}, i, j = 1, 2, i \neq j. \end{cases}$$

This definition is illustrated in Figure 7.2. The definition may look unnecessary complex, but is used to make the definition of the bounding area simple.

The half-plane that contains point c and whose boundary passes through a_1 and a_2 is denoted by $H(a_1, a_2; c)$. A half-plane can be considered as a disc with a radius of $-\infty$.

The bounding area, called flintstone, that is the basis of the approximation algorithm is defined in terms of discs with signed radii and the $\pm$-operator:

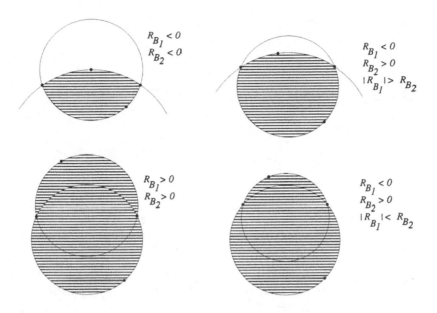

Figure 7.2. $B_1 \pm B_2$.

DEFINITION 7.3 (FLINTSTONE) *Let* $P = v_p \ldots v_s$, *and let* v_q *be a vertex of P such that* $D(v_p, v_s; v_q)$ *contains all vertices lying in* $H(v_p, v_s; v_q)$. *If* $P \subset H(v_p, v_s; v_q)$, *then the flintstone F of P is defined as* $F(P) = D(v_p, v_s; v_q) \cap H(v_p, v_s; v_q)$, *otherwise* $F(P) = D(v_p, v_s; v_q) + D(v_p, v_s; v_r)$, *where* v_r *is a vertex of P not lying in* $H(v_p, v_s; v_q)$ *and such that* $D(v_p, v_s; v_r)$ *contains all vertices in* $H(v_p, v_s; v_r)$.

Informally, the flintstone is the smallest intersection or union of two discs touching v_p and v_s that contains $v_p \ldots v_s$. The definition is illustrated in Figure 7.3 for the case that the flintstone is an intersection of two discs or a disc and a half-plane. It is easily verified that $P \subset F(P)$, so that $F(P)$ is indeed a bounding area for P.

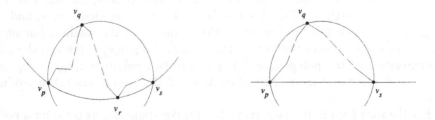

Figure 7.3. 2D flintstone. Left: $F(P) = D(v_p, v_s; v_q) \pm D(v_p, v_s; v_r)$. Right: $F(P) = D(v_p, v_s; v_q) \cap H(v_p, v_s; v_q)$.

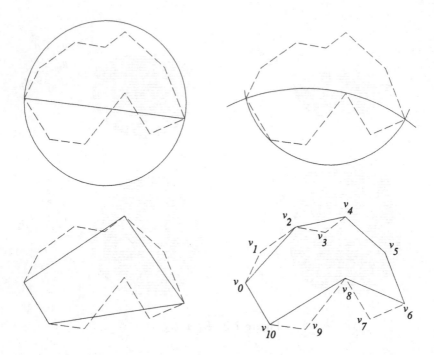

Figure 7.4. Top left: smallest bounding disc and zeroth order approximation. Top right: flintstone of $v_6 \ldots v_0$. Bottom: first and second order approximation.

Note that such a vertex v_q always exists. After all, there exists a disc touching v_p and v_s that contains all vertices of P in $H(v_p, v_s; v_i)$ for some $p < i < s$. The smallest possible one touches at least one vertex v_j, $p < j < s$. We call one of these vertices v_q. Symmetrically, if not all vertices of P lie in $H(v_p, v_s; v_q)$, then also the v_r in the definition exists.

7.2.2 Approximation

The hierarchical approximation of a closed polygon of N_v vertices starts with the calculation of the smallest bounding disc (SBD), that is, the smallest disc that contains all vertices. This disc touches at least two vertices, say v_i and v_j, $i < j$. If more than two vertices lie on the boundary of the smallest bounding disc, we take two vertices that are farthest apart. Edge $v_i v_j$ is the zeroth order approximation of the polygon, dividing it into two polylines $v_i v_{i+1} \ldots v_j$ and $v_j v_{j+1} \ldots v_i$ (here and in the rest of Section 7.2 the indices are taken modulo N_v).

For the next level in the hierarchy of approximations, let us consider a polyline $P = v_p \ldots v_s$. The approximation depends on the flintstone $F(P)$. If $F(P) = D(v_p, v_s; v_q) \cap H(v_p, v_s; v_q)$, then P is approximated by the edges $v_p v_q$ and $v_q v_s$. If $F(P) = D(v_p, v_s; v_q) \perp D(v_p, v_s; v_r)$ and if the radius of $D(v_p, v_s; v_q)$ is larger than the radius of $D(v_p, v_s; v_r)$, then P is approximated by the edges $v_p v_q$

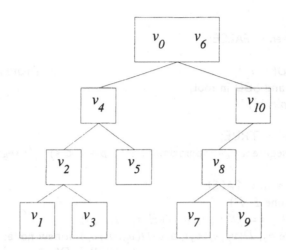

Figure 7.5. Flintstone tree of example in Figure 7.4.

and $v_q v_s$, otherwise by $v_p v_r$ and $v_r v_s$. Notice that the radii of $D(v_p, v_s; v_q)$ and $D(v_p, v_s; v_r)$ are signed.

An edge $v_p v_s$ is not subdivided if $p + 1 = s$, in which case this edge is in the original polygon. In order to construct the complete hierarchy, the iteration is performed until all vertices are contained in the approximation polygon, which then coincides with the original one. By definition, an edge $v_p v_{p+1}$ has no flintstone. Figure 7.4 gives a simple example of a polygon and the approximation of level zero, one, and two.

If there exists a disc touching v_p and v_s that contains $v_{p+1} \ldots v_{s-1}$, then the '$\pm$' in the definition of the flintstone is a '$\cap$'. By construction of the approximation, each approximated polyline is contained in a disc. So, all flintstones are the *intersection* of two discs or a disc and a half-plane. In particular the degenerate case that a flintstone is the union of a half-plane and a disc simply cannot occur. That would be the case if a vertex v_i, $p < i < s$, lies on the line through v_p and v_s and outside the line segment $v_p v_s$, but by construction of the approximation, this never happens. The shape of the intersection of two discs gave rise to the name 'flintstone'.

A binary tree is a natural data structure to store the hierarchical approximation. The root, level zero of the tree, contains vertices v_i and v_j. The left subtree stores vertices $v_{i+1}, \ldots, v_{j-1}$ such that the symmetric or infix order traversal yields the successive vertices of the polygon. The right subtree stores vertices $v_{j+1}, \ldots, v_{i-1}$ analogously. A level-ℓ approximation simply corresponds to the levels $0, \ldots, \ell$ of the tree. Figure 7.5 shows the complete hierarchy tree of the example in Figure 7.4.

The bounding areas are stored as follows. The root stores the smallest bounding disc. A node containing vertex v_q at level ℓ, $\ell \geq 1$, of the tree, contains $D(v_p, v_s; v_q)$ and $D(v_p, v_s; v_r)$, where v_p is the predecessor and v_s the successor of v_q at approximation level ℓ. For example, the node containing v_2 in Figure 7.5

```
Flintstones2 ()
{ bool Completed = FALSE;

  compute SBD(v_i, v_j);                                    // smallest bounding disc
  store v_i, v_j, and SBD in root;
  while (!Completed)
  {
    Completed = TRUE;
    for (each segment v_p v_s approximating P, p+1 < s)   // segment to be split
    {
      Completed = FALSE;
      determine F(P);
      if (F(P) == D(v_p, v_s; v_q) ∩ H(v_p, v_s; v_q))
        store v_q, D(v_p, v_s; v_q), and D(v_p, v_s; v_r) in new node;
      else                              // F(P) is D(v_p, v_s; v_q) ± D(v_p, v_s; v_r)
        if ( R(D(v_p, v_s; v_q)) > R(D(v_p, v_s; v_r)) )
          store v_q, D(v_p, v_s; v_q), and D(v_p, v_s; v_r) in new node;
        else
          store v_r, D(v_p, v_s; v_q), and D(v_p, v_s; v_r) in new node;
    }
  }
}
```

Algorithm 7.1. 2D flintstone tree construction algorithm.

stores $D(v_0, v_4; v_2)$ and $D(v_0, v_4; v_3)$.

I shall call the approximation and localization scheme 'flintstone scheme', and the tree that stores the hierarchy of approximations and bounding volumes a 'flintstone tree', also in 3D.

The algorithm to construct the complete flintstone tree is summarized in Algorithm 7.1 in pseudo C-language code. The time complexities for the best and worst case and the storage complexity are given by the following theorems.

THEOREM 7.1 (TIME COMPLEXITY) *The best case time complexity to construct the complete flintstone tree in 2D is* $\Theta(N_v \log N_v)$. *The worst case complexity in 2D is* $\mathcal{O}(N_v^2)$.

Proof. The smallest bounding disc can be found in $\Theta(N_v)$ time [Megiddo, 83]. Two vertices on the boundary of the disc that are farthest apart can be found in $\Theta(N_v \log N_v)$ [Preparata and Shamos, 85]. Finding the new vertex to be included in the approximation polygon can be done in $\mathcal{O}(n)$ time for a polyline of n segments, by testing the vertices sequentially.

In the best case, the polygon is split into equally sized parts, giving $\lceil \log N_v \rceil$ iterations. The i-th iteration then treats 2^i polylines of size $N_v/2^i$. The best case time complexity is thus

$$\Theta(N_v) + \Theta\left(\sum_{i=0}^{\lceil \log N_v \rceil} 2^i (N_v/2^i)\right) = \Theta(N_v \log N_v).$$

In the worst case a polyline of n segments is split into one polyline of size one and another of size $n-1$. The worst case time complexity is therefore

$$\Theta(N_v) + \mathcal{O}(\sum_{i=1}^{N_v} i) = \mathcal{O}(N_v^2). \quad \square$$

THEOREM 7.2 (STORAGE COMPLEXITY) *The storage complexity of the complete flintstone tree in 2D is* $\Theta(N_v)$.

Proof. The flintstone tree stores each vertex exactly once. The root stores two vertices, all other nodes store one vertex, so that there are $N_v - 1$ nodes. The flintstone tree thus requires $\Theta(N_v)$ storage space. $\square$

7.2.3 Adaptive approximation

The approximation algorithm described in the previous section replaces each edge in the current approximation by two new edges, unless the edge cannot be refined. One can apply this procedure a fixed number of times, or stop the iteration when the approximation polygon is within a specified error bound with respect to a chosen criterion. However, the error of one part of the approximation polygon can be very different from another part. This is clearly visible in Figure 7.6, showing the original polygon (the outer border of the mainland of The Netherlands, Belgium, and Luxembourg) with the zero-order level of approximation, and two other levels of approximation. Some parts of the approximation polygon exhibit much more detail than other parts. The approximation algorithm can overcome this unevenness by only refining an edge if the error of *that edge* (rather than the whole polygon) is larger than a given bound. Such an algorithm is called adaptive.

In principle, any error criterion can be used. An approximation error tailored to a covering by flintstones is based on the two widths associated with a flintstone (illustrated in Figure 7.7):

DEFINITION 7.4 (WIDTHS OF FLINTSTONE) *Let* $P = v_p \dots v_s$ *have an associated flintstone* $F(P) = D(v_p, v_s; v_q) \cap D(v_p, v_s; v_r)$ *or* $F(P) = D(v_p, v_s; v_q) \cap H(v_p, v_s; v_q)$. *Let* m_1 *be the point on* $D(v_p, v_s; v_q)$ *at the same side of edge* $v_p v_s$ *as* v_q *that has the largest distance* w_1 *to* $v_p v_s$. *Let* c_1 *be the center of* $D(v_p, v_s; v_q)$, R_1 *its radius,* $r_1 = d(v_p, v_s)/2$, *and* h_1 *the distance between* c_1 *and* $v_p v_s$, *i.e.* $h_1 = d(v_p, v_s)/2$. *If* $R_1 > 0$, *then* $w_1 = R_1 + h_1$, *if* $R < 0$, *then* $w_1 = -(R_1 + h_1)$. *If* $F(P) = D(v_p, v_s; v_q) \cap D(v_p, v_s; v_r)$, *then* w_2 *is analogously defined with respect to* $D(v_p, v_s; v_r)$, *and if* $F(P) = D(v_p, v_s; v_q) \cap H(v_p, v_s; v_q)$, *then* $w_2 = 0$.

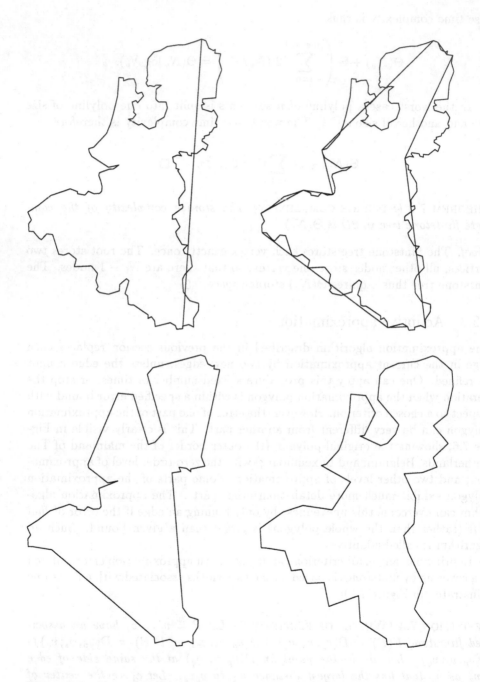

Figure 7.6. Top row: original polygon together with the zeroth order (left) and fourth order (right) approximation. Bottom row: higher order approximation of 59 edges (left) and adaptive approximation of 60 edges (right).

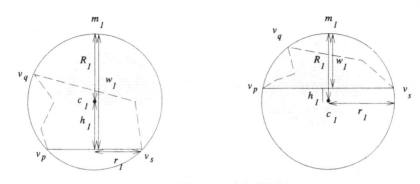

Figure 7.7. Width $w_1 = R_1 + h_1$ if $R_1 > 0$ (left), $w_1 = -(R_1 + h_1)$ if $R_1 < 0$ (right).

Note that the zero width associated with $H(v_p, v_s; v_q)$ agrees with the interpretation that the half-plane is a disc with a radius of $-\infty$. Verbally, the widths of a flintstone are the largest distances between the boundaries of the discs and the line through v_p and v_s.

The covering approximation error of a single flintstone, say flintstone i, is now defined as $\max\{w_{i,1}, w_{i,2}\}$. The error of approximation of a polyline by m flintstones is

(e5)
$$\max_{1 \leq i \leq m} \max\{w_1, w_2\}.$$

The adaptive approximation in Figure 7.6 is constructed using this error criterion.

The tree representation of an adaptive approximation is a subgraph of the tree of the non-adaptive approximation of the same level. An adaptive approximation polygon generally consists of fewer edges than a non-adaptive one of the same level of approximation. Alternatively, an adaptive approximation polygon generally is of a higher approximation level than a non-adaptive one of about the same number of edges, usually resulting in a better approximation, i.e. a smaller approximation error. This is illustrated in Figure 7.6 at the bottom row: the adaptive approximation and the non-adaptive one have about the same number of edges, but the adaptive approximation has a smaller error and is of a higher approximation level.

7.3   Flintstones in 3D

In this section we consider the approximation and localization of a 3D closed simple polyhedron of genus zero, that is, without holes. The approximation algorithm is based on the way the localization of a part of the closed polyhedron, which is an open polyhedron, is performed.

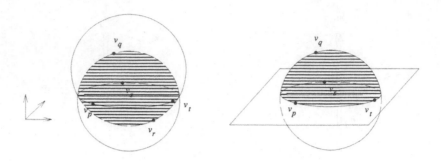

Figure 7.8. 3D flintstones.

7.3.1   Localization

The definition of the bounding volume is based on the notion of a ball with a signed radius. $B(a_1, a_2, a_3; b)$ denotes the ball touching a_1, a_2, a_3, b with a signed radius, defined analogously to Definition 7.1. The $\pm$-operator for balls with a signed radius is precisely defined by Definition 7.2. The half-space containing c and whose boundary passes through a_1, a_2, and a_3, is denoted by $H(a_1, a_2, a_3; c)$. A half-space can be considered as a ball with a radius of $-\infty$. Before presenting the approximation algorithm, the basic bounding volume, flintstone, must be defined:

DEFINITION 7.5 (FLINTSTONE) *Let P be an open polyhedron, and v_p, v_s, and v_t three distinct vertices on the closed boundary of P. Let v_q be a vertex of P such that $B(v_p, v_s, v_t; v_q)$ contains all vertices lying in $H(v_p, v_s, v_t; v_q)$. If $P \subset H(v_p, v_s, v_t; v_q)$, then the flintstone F of P is defined as $F(P) = B(v_p, v_s, v_t; v_q) \cap H(v_p, v_s, v_t; v_q)$, otherwise $F(P) = B(v_p, v_s, v_t; v_q) \pm B(v_p, v_s, v_t; v_r)$, where v_r is a vertex of P not in $H(v_p, v_s, v_t; v_q)$ such that $B(v_p, v_s, v_t; v_r)$ contains all vertices in $H(v_p, v_s, v_t; v_r)$.*

Note that, analogous to the 2D situation, such a v_q always exists, and also a v_r if not all vertices of P lie in $H(v_p, v_s, v_t; v_q)$. This definition is illustrated in Figure 7.8 for the case that the flintstone is an intersection of two balls or a ball and a half-space. It is easily verified that $P \subset F(P)$, so $F(P)$ is indeed a bounding volume for P.

7.3.2   Approximation

The hierarchical approximation of a closed polyhedron starts with the calculation of the smallest ball that touches at least three vertices, say v_i, v_j, and v_k, and contains all vertices. Such a ball always exists, but need not be the smallest bounding ball, which may touch only two vertices.

The three shortest paths of edges in the polyhedron running between v_i and v_j, v_j and v_k, and between v_k and v_i, divide the closed polyhedron into two

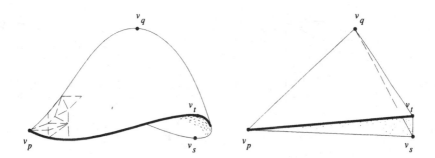

Figure 7.9. Triangle $v_p v_s v_t$ is split into three at a point inside the polyhedral surface.

open polyhedra. Triangle $v_i v_j v_k$ is the zeroth order approximation of the closed polyhedron.

At all next levels of the hierarchical approximation we consider an open polyhedron P of more than three vertices, and three distinct vertices v_p, v_s, and v_t on the boundary of P. If $F(P) = B(v_p, v_s, v_t; v_q) \cap H(v_p, v_s, v_t; v_q)$, then P is approximated by the three triangles $v_p v_q v_s$, $v_q v_s v_t$, and $v_p v_q v_t$. If $F(P) = B(v_p, v_s, v_t; v_q) \pm B(v_p, v_s, v_t; v_r)$ and if the (signed) radius of $B(v_p, v_s, v_t; v_q)$ is larger than the radius of $B(v_p, v_s, v_t; v_r)$, P is approximated by $v_p v_q v_s$, $v_q v_s v_t$, and $v_p v_q v_t$, as illustrated in Figure 7.9. Otherwise P is approximated by $v_p v_r v_s$, $v_r v_s v_t$, and $v_p v_r v_t$.

At all levels of approximation there is an open polyhedral part of the original polyhedron associated with each approximation triangle $v_p v_s v_t$. The boundary polygon of that open polyhedron consists of the three shortest paths between v_p, v_s, and v_t. Note that each pair of these paths may partly coincide, especially near v_p, v_s, or v_t.

If the triangles are always split into three new ones as described above, the new triangles become more and more elongated, because each time the angles at the vertices are divided. Moreover, the edges are retained in all next levels, which

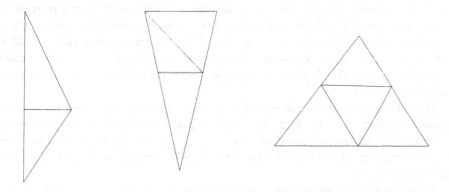

Figure 7.10. Schematic splitting at sides.

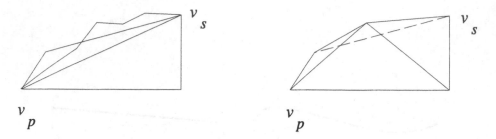

Figure 7.11. Two neighboring triangles are split at their common side.

will generally not lead to a good approximation. To avoid too thin triangles, the angle is not split if it is less than some chosen value. In that case the triangle is split at two sides, as in Figure 7.10 (middle). A triangle can also have two angles that are too small to be split, in which case the triangle is split at one side, as in Figure 7.10 (left). If a triangle is split at a side, its neighboring triangle sharing that side should also be split at the same point, in order to prevent cracks in the approximation polyhedron. A triangle can thus be forced to split, even if it has no small angles. Therefore, a triangle can also split at all three sides, into four new triangles as illustrated in Figure 7.10 (right).

Because a triangle can be forced to split at a side by its neighbors, the effect of splitting at a side propagates through the current approximation polyhedron. Therefore, all edges in the current approximation polyhedron that must be split should be determined first. Only then it is known how the splitting of all triangles should be done.

The position where a side should be split is at a vertex on the shortest path between the end points, which is part of the boundary polygon of the open polyhedron that is approximated by the triangle. The vertex where the split is performed is the one that gives the smallest difference between the lengths of the new sides. If the shortest path consists of a single edge, there is no such vertex, and the split is not performed. So, no new vertices are introduced. Such a split of a triangle into two new ones, together with its neighboring triangle, is shown in Figure 7.11. As stated before, two paths may partly coincide, which may affect the result of the splitting. For if the vertex where a side is split is part of two paths, then degenerate triangles will result. This, however, causes no problem, since the degenerate triangles can simply be recognized by their coalescing vertices.

The refinement iteration is repeated until no approximated open polyhedron contains interior vertices, and the shortest paths between the vertices of all approximation triangles consist of single edges. This is the most detailed level of approximation. By definition, the approximation triangles at the lowest level have no flintstone.

At an almost final approximation level, the approximation may locally be not very accurate. Look for example at Figure 7.12 (note that the dashed lines do

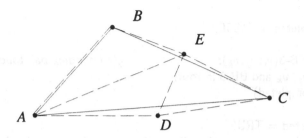

Figure 7.12. Triangle ABC approximates original triangles, drawn with dashed lines.

not denote the way of triangle refinement, but represent the original triangles).
A possible sequence of splittings is represented in the following table, which lists
for each of the successive approximating triangles: (i) the shortest paths between
its vertices, (ii) the vertices and original triangles that it approximates, and (iii)
into which triangles it is refined.

triangle	shortest paths	approximates	split into
ABC	AB, BEC, CDA	vert: A, B, C, D, E tr: ABE, ADE, CDE	ABD, BCD
ABD	AB, BED, DA	vert: A, B, D, E tr: ABE, ADE	ABE, ADE
BCD	BEC, CD, DEB	vert: B, C, D, E tr: CDE	CDE, BDE
BDE	BE, ED, DEB		

So triangle BCD is split into CDE and BDE, but the latter one approximates
a void part of the polyhedron. Triangles that do not approximate a part of the
polyhedron have coalescing shortest paths between their vertices, which is easily
tested. Such triangles are discarded from the approximation. In that case, an
approximation triangle can be refined into only one new triangle, while in the
normal case it has two to four children.

Because no new vertices are introduced when splitting a triangle at a side,
the shortest paths consist of edges of the original polyhedron. And because
triangles that do not approximate a part of the polyhedron are discarded, all
the final approximating triangles coincide with the original triangles. So, at the
most detailed level of approximation the original polyhedron is recovered.

In order to avoid inaccurate approximations at an almost final level, like
illustrated above, the split procedure must take the triangulation topology into
account, not only the geometry.

If there exists a ball touching v_p, v_s, and v_t and containing P, then the '$\pm$'
in the definition of the flintstone is a '$\cap$'. If no sides of this triangle are split,
the open polyhedra associated with the new triangles are also contained in a
single ball. However, if one or more sides are split, these polyhedra need not
be contained in a single ball. So, unlike the 2D case, flintstones in 3D are not
always the intersection of two balls or a ball and a half-space, but may also be

```
Flintstones3 ()
{  bool Completed = FALSE;

    compute BB-3(v_i, v_j, v_k);                    // bounding ball touching 3 vertices
    store v_i, v_j, v_k and BB-3 in root;
    while (!Completed)
    {
        Completed = TRUE;
        for (each v_p v_s v_t approximating a P of more than 3 vertices)
        {
            Completed = FALSE;
            determine the sides to split;
        }
        for (each v_p v_s v_t approximating a P of more than 3 vertices)
        {
            split triangle;
            divide P accordingly;
            create children of v_p v_s v_t and store new triangles and their flintstones;
        }
    }
}
```

Algorithm 7.2. 3D flintstone tree construction algorithm.

the union of two balls. As a result, the flintstone can degenerate to the union of a half-space and a ball. That will be the case when a vertex v_i of the approximated open polyhedron lies in the plane through the vertices v_p, v_s, and v_t, and outside the triangle $v_p v_s v_t$. In many data sets obtained from experimental applications no four vertices are coplanar, so that this problem does not arise. By contrast, many synthetic data sets contain groups of four coplanar vertices, such as the candlestick object in Figure 5.5 and the bottle in Figure 5.10.

In 2D the flintstone tree stores vertices, while the edges of successive approximations are implicitly defined. There is no simple analogous scheme in 3D that implicitly represents the triangles. They are therefore stored explicitly. The root contains the center and radius of the ball bounding the whole polyhedron. The two sons of the root contain v_i, v_j and v_k, approximating the open polyhedra P_1 and P_2, and the centers and signed radii of the balls that define the flintstones. All nodes except the root can have up to four sons, which are constructed as described above.

The algorithm to build the complete flintstone tree is summarized in Algorithm 7.2 in pseudo C-language code. The time complexities of the algorithm for the best and worst case, and the storage complexity are given by the following theorems.

THEOREM 7.3 (TIME COMPLEXITY) *The best case time complexity to construct the complete flintstone tree in 3D is $\mathcal{O}(N_v(\log N_v)^2)$. The worst case complexity in 3D is $\mathcal{O}(N_v^2 \log N_v)$.*

Proof. The smallest bounding ball is found in $\Theta(N_v)$ time [Megiddo, 83]. In 3D, this ball may touch only two vertices. Finding a third vertex such that the ball touching these three vertices contains them all, may take another $\mathcal{O}(N_v)$ time. Finding the new vertex to be included in the approximate polyhedron can be done in $\mathcal{O}(n)$ time for a polyhedron of n vertices. Both operations can simply be performed by successively testing all candidates. Finding a shortest path in a polyhedron of n vertices takes $\Theta(n \log n)$ time [Dijkstra, 59].

In the best case, the open polyhedron is split into equally sized parts, giving $\Theta(\log N_v)$ iterations. A polyhedron is split into at most four parts, but the order of complexity is not affected if we let the total number of polyhedra at the i-th iteration be 2^i. The best case complexity is therefore

$$\Theta(N_v) + \mathcal{O}\left(\sum_{i=0}^{\lceil \log N_v \rceil} 2^i((N_v/2^i)\log(N_v/2^i))\right) = \mathcal{O}\left(\sum_{i=0}^{\lceil \log N_v \rceil} N_v((\log N_v) - i)\right)$$

$$= \mathcal{O}\left(N_v \sum_{i=0}^{\lceil \log N_v \rceil} i\right) = \mathcal{O}(N_v(\log N_v)^2).$$

In the worst case the number of vertices of the polyhedron to be approximated decreases by one at each iteration. The worst case time complexity is therefore

$$\Theta(N_v) + \mathcal{O}\left(\sum_{i=1}^{N_v} i \log i\right) = \mathcal{O}\left(\sum_{i=N_v/2}^{N_v} i \log i\right) = \mathcal{O}(N_v^2 \log N_v). \quad \square$$

THEOREM 7.4 (STORAGE COMPLEXITY) *The storage complexity of the complete flintstone tree in 3D is $\Theta(N_t)$.*

Proof. Almost all internal nodes have two to four children, only at the lowest levels triangles are possibly refined into one new triangle, so that only $\mathcal{O}(N_v)$ nodes have one child. Because the leaves of the tree contain the original N_t triangles, there are $\Theta(N_t)$ internal nodes. The total storage space is thus $\Theta(N_t)$. $\quad \square$

7.3.3 Adaptive approximation

The approximation error of a flintstone covering in 3D is again Error (e5), i.e. $\max\{w_1, w_2\}$, and the widths are defined in complete analogy to the 2D case:

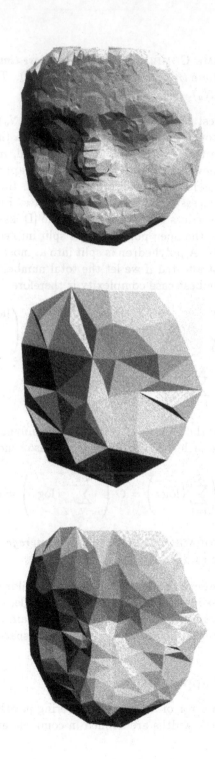

Figure 7.13. Top: polyhedral object of 2930 triangles (from Figure 5.6). Middle: adaptive flintstone approximation of 89 triangles. Bottom: adaptive flintstone approximation of 277 triangles.

DEFINITION 7.6 (WIDTHS OF FLINTSTONE) *Let an open polyhedron P have a flintstone $F(P) = B(v_p, v_s, v_t; v_q) \cap B(v_p, v_s, v_t; v_r)$ or $F(P) = B(v_p, v_s, v_t; v_q) \cap H(v_p, v_s, v_t; v_q)$. Let m_1 be the point on $B(v_p, v_s, v_t; v_q)$ at the same side of segment $v_p v_s v_t$ as v_q that has the largest distance w_1 to $v_p v_s v_t$. Let c_1 be the center of $B(v_p, v_s v_t; v_q)$, R_1 its radius, $r_1 = r(v_p, v_s, v_t)$, and h_1 the distance between c_1 and $v_p v_s v_t$. If $R_1 > 0$, then $w_1 = R_1 + h_1$, if $R < 0$, then $w_1 = -(R_1 + h_1)$. If $F(P) = B(v_p, v_s, v_t; v_q) \cap B(v_p, v_s, v_t; v_r)$, then w_2 is analogously defined with respect to $B(v_p, v_s, v_t; v_r)$, and if $F(P) = B(v_p, v_s, v_t; v_q) \cap H(v_p, v_s, v_t; v_q)$, then $w_2 = 0$.*

The $r(v_p, v_s, v_t)$ in this definition is the radius of the disc touching v_p, v_s, v_t, as in the definition of the γ-neighborhood, see page 24. The zero width associated with $H(v_p, v_s, v_t; v_q)$ agrees with the interpretation that the half-space is a ball with a radius of $-\infty$.

Figure 7.13 shows an example of a polyhedral object and two adaptive flintstone approximations using this error criterion.

7.4 Hierarchical operations

Section 6.1 described the use of a hierarchical approximation and localization for an efficient point location test: if the query point is not inside the flintstone of an approximation face, the test is not affected by the replacement of that part of the polygon/hedron by the approximation. Testing whether a point X is internal to a flintstone amounts to calculating the distances between X and the centers of the balls. If F is the intersection of two balls and both distances are less than the corresponding radii of the balls, X is internal to the flintstone. If F is the union of two balls (in 3D only), and one of the distances is less than the radius of the corresponding ball, X lies inside the flintstone.

Another operation that can efficiently be performed hierarchically is the intersection of two polygons or polyhedra. Detection and computation of polygonal or polyhedral intersections are fundamental problems in hidden surface elimination, motion planning, and linear programming. The flintstone scheme can be exploited as follows. The hierarchical intersection algorithm first tests if the bounding balls at the roots intersect. If they don't, then neither do the objects themselves. Otherwise the algorithm proceeds with the next level of the deepest subtree and the same level of the other (sub)tree, testing all pairs of bounding volumes for intersection. This process locally continues for those pairs that intersect, and stops for those that don't. When flintstones at the lowest level are tested and found to intersect, the original boundary faces themselves are tested.

If flintstone F is defined by balls B_1 and B_2, and flintstone G by balls C_1 and C_2, the following combinations can occur:

1. $F = B_1 \cap B_2$, $G = C_1 \cap C_2$,
2. $F = B_1 \cup B_2$, $G = C_1 \cup C_2$,
3. $F = B_1 \cap B_2$, $G = C_1 \cup C_2$.

In these situations, F and G intersect in the following corresponding situations:

1. $B_1 \cap C_1 \neq \phi \wedge B_2 \cap C_1 \neq \phi \wedge B_1 \cap C_2 \neq \phi \wedge B_2 \cap C_2 \neq \phi$,

2. $B_1 \cap C_1 \neq \phi \vee B_2 \cap C_1 \neq \phi \vee B_1 \cap C_2 \neq \phi \vee B_2 \cap C_2 \neq \phi$,

3. $(B_1 \cap C_1 \neq \phi \wedge B_2 \cap C_1 \neq \phi) \vee (B_1 \cap C_2 \neq \phi \wedge B_2 \cap C_2 \neq \phi)$.

For 2D balls (discs), only situation 1 can occur.

Testing whether two balls intersect amounts to comparing the distance between their centers with the sum of the radii. If the distance is larger, they do not intersect, otherwise they do intersect.

For intersection detection, the algorithm can stop as soon as a single intersection is detected. In order to actually compute the intersection, the iteration must be continued until all intersecting boundary faces are found, and then the actual intersection must be computed. After computing the intersections of the polygons or polyhedra, the solid object that is the intersection, union, or difference of the two objects can be determined in a manner similar to the methods described by [Günther, 88] and [Ponce and Faugeras, 87].

For both the point location and the intersection algorithm, the basic operation is the computation of the distance between two points. This is very simple and computationally cheap. Especially in 3D this is very efficient compared to other localization schemes. For example a point-in-prism test needs to consider five faces, and a prism–prism intersection requires twenty-five polygon–polygon intersections.

In 2D, the localization by ellipses in the arc-tree scheme was found to be efficient, compared to other schemes [Dominguez and Günther, 91]. The point-in-flintstone test is as cheap as a point-in-ellipse test, which requires the distances of the point to the two focal points of the ellipse. However, the intersection of two ellipses is not as simple as the intersection of two flintstones. It should be noted though, that the performance of such hierarchical operations not only depends on the efficiency of calculations with a single bounding volume, but also on the quality of localization for the whole object.

7.5   Concluding remarks

The flintstone scheme has some advantages over other boundary-based schemes: it is hierarchical, an approximation as well as a localization scheme, and the bounding volumes are efficient in storage and for computations.

Clearly, the flintstone of a polyline $v_p \ldots v_s$ is a γ-neighborhood of v_p and v_s for two appropriate parameters, and the flintstone of an open polyhedron is a γ-neighborhood of three vertices v_p, v_s, and v_t, for two appropriate parameters. However, the purpose of a flintstone is to provide a bounding area or volume containing all vertices in the polygon or polyhedron, while the purpose of the γ-neighborhood is to test emptiness of that area or volume.

In interactive applications, vertices may be dynamically added to the polygon or polyhedron that must be approximated. The flintstone approximation must

then be dynamically adapted by traversing the flintstone tree and finding the
first flintstone that does not anymore contain the new vertex. The subtree rooted
by the corresponding node must then be rebuilt.

Even if the original object is a simple polygon or polyhedron (informally, the
boundary does not intersect itself), the flintstone approximation need not be
simple. Especially at the first few approximation levels the polygon or polyhedron may cut itself. At higher levels of approximation, however, this will rarely
happen. All approximation schemes in Section 6.3 except the Delaunay pyramid, which can only be used for $2\frac{1}{2}$D polyhedra, suffer from the same problem.
A future research direction is the development of a hierarchical approximation
and localization scheme that always yields simple approximation polygons or
polyhedra when the original is simple.

The time complexity for the flintstone tree construction is the same as for
many of the schemes described in Section 6.3. Testing for point inclusion and
flintstone intersection is done by calculating distances between points, which is
computationally cheaper than the calculations on bounding volumes of many
other schemes. For simplicity and efficiency of storage and computations a half-
space in the definition of a flintstone can be represented by a ball of large radius.
In order to compare the efficiency of the flintstone scheme with the efficiency of
other methods, those should also be implemented and applied to the same data
sets. Such a comparison is done by [Dominguez and Günther, 91] between the
arc tree, the strip tree, and approximation and localization by the Bézier scheme
(introduced in the next chapter). Extending the comparison with the flintstone
scheme is a possible subject of future research.

8

Smooth curves and surfaces

This chapter gives an introduction to spline curves and surfaces, the Bézier formulation for these, and the notion of geometric continuity. Some consequences of geometric continuity regarding visual aspects are then derived.

8.1 Introduction

In the previous chapters the constructed object boundaries are all piecewise linear, i.e. a polygon in 2D and a closed polyhedron of triangles in 3D. For esthetic purposes or physical requirements a smooth boundary is often desired. An inherently smooth boundary that has a continuously changing tangent line or plane cannot consist of linear segments, but must consist of curved pieces: curve segments for a boundary in 2D, and surface segments, or patches, for a boundary in 3D. One particularly useful way of representing curves and surfaces is the Bézier formulation, because of the geometrically meaningful definition. We will use Bézier curves and surfaces for the construction of smooth boundaries. There are several types of smoothness, or continuity, for curves and surfaces, and this chapter introduces some of them.

Section 8.2 treats representations of curves, in particular Bézier curves, and notions of continuity of curves. Section 8.3 does the same for surfaces. Some visual aspects of continuity are discussed in Section 8.4. A more elaborated treatment of these subjects is given by [Veltkamp, 92d].

8.2   Curves

8.2.1   Fundamentals

A curve in *functional* form is a scalar function $f : \mathbb{R} \to \mathbb{R}$. A curve in *parametric* form is a vector-valued function $P : \mathbb{R} \to \mathbb{R}^k$, $k \geq 2$, each coordinate component being a function of the parameter: $P(t) = (P^1(t), \ldots, P^k(t))^T$, $t_{min} \leq t \leq t_{max}$, for a curve in kD. Parametric curves have some advantages over functional curves, for example they are independent of a particular coordinate frame. Another advantage is their capability to represent closed curves, which is important for our purpose, the construction of closed boundaries. In the following we only deal with parametric curves.

One particular type of parametric curve is the *polynomial* curve:

$$(8.1) \qquad\qquad P(t) = p_0 + p_1 t + p_2 t^2 + \ldots + p_n t^n,$$

for some finite integer $n \geq 0$, the degree of the polynomial curve, and $p_0, \ldots, p_n \in \mathbb{R}$. The coefficients p_i are vectors whose coordinate components are the coefficients for each coordinate function. The number of coefficient vectors, $n + 1$, is the *order* of the polynomial. Polynomial curves are often used because they are easy to handle, for example for evaluation and the calculation of derivatives.

In Equation (8.1) the curve is represented as a linear combination of the so-called power basis functions $1, t^1, \ldots, t^n$. We can also use other polynomial basis functions $B_i^n(t)$:

$$(8.2) \qquad\qquad P(t) = \sum_{i=0}^{n} p_i B_i^n(t).$$

The B_i^n are often called blending functions and the p_i weights or control points.

The derivative of a curve is a vector, the derivatives are taken componentwise. The i-th derivative of a curve $P(t)$ is denoted $P^{(i)}(t)$. To avoid possible problems with the parameterization of the curve, we assume in the rest of this thesis that the first derivative vector of all curves is not equal to the null-vector: $P^{(1)}(t) \neq 0$. Such a curve and its parameterization are called *regular*.

A *piecewise* polynomial curve is defined segment by segment. The parameter range is then partitioned into subranges: $t_{min} = t_0 < t_1 \leq \ldots t_m = t_{max}$, where t_i are fixed parameter values. The curve is defined for each subrange: $P(t) = P_i(t)$, for $t_i \leq t < t_{i+1}$. We are usually interested in positionally continuous piecewise curves:

$$\lim_{t \uparrow t_i} P_i(t) = P_{i+1}(t_i),$$

as in Figure 8.1, but that is not always necessary.

The parameter subranges can be transformed so as to provide a local parameter u, $u_{min} \leq u \leq u_{max}$, for example the normalized range $[0, 1]$. $u = (t - t_i)/(t_{i+1} - t_i)$, $t_{i+1} \neq t_i$. The curve $P_i(t)$ can then be *reparameterized* into $\tilde{P}_i(u) = P_i(t(u))$, $u_{min} \leq u \leq u_{max}$. In the previous example

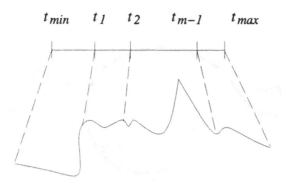

t_{min} t_1 t_2 t_{m-1} t_{max}

Figure 8.1. Piecewise curve.

$t(u) = t_i + u(t_{i+1} - t_i)$. In the general case, the local parameter ranges are independent of each other, so the ranges may be disconnected or overlap.

The word *spline* is an East Anglian dialect word, denoting a metal or wooden strip, bent around pins to form an esthetically pleasing shape. It was observed that under gentle bending the shape corresponds to a piecewise *cubic* polynomial function having continuous first and second derivatives. In the context of mathematical curves a spline can be of any polynomial degree. The theory of splines originates from approximation theory. Spline approximation in its present form first appeared in a paper by Schoenberg, who developed methods for the smooth approximation of empirical tables [Schoenberg, 46]. In approximation theory a spline of order $n + 1$ is generally defined as a piecewise polynomial of degree n that is everywhere C^{n-1}-continuous (see Section 8.2.2 for C^n-continuity). In geometric modeling it is sometimes desired to model discontinuities on purpose, so that the continuity requirement in the definition is left out:

DEFINITION 8.1 (POLYNOMIAL SPLINE) *A polynomial spline function is a piecewise polynomial function, and a polynomial spline curve is a curve whose components are polynomial spline functions.*

We are interested in the continuity of spline curve segments $P(u)$, $u_{min} \leq u \leq u_{max}$ and $Q(w)$, $w_{min} \leq w \leq w_{max}$, at points $P(u_0)$ and $Q(w_0)$ on the curves, in particular at the end points $P(u_{max})$ and $Q(w_{min})$.

8.2.2 Continuity

Parametric continuity is the classical notion of continuity of functions in analysis: if a function is n times continuously differentiable, or more exactly, the derivatives exist and are continuous, then the function is n-th order parametric continuous. Applying this concept to curves, we get the following definition:

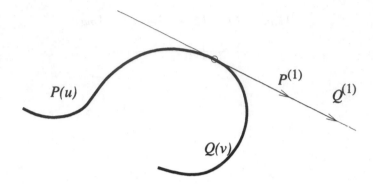

Figure 8.2. Two curve segments joining with derivatives of same direction but different magnitude.

DEFINITION 8.2 (PARAMETRIC CONTINUITY) *Two curves $P(u)$ and $Q(w)$ are n-th order parametric continuous, $n \geq 0$, at u_0 and w_0, if and only if $P^{(i)}(u_0) = Q^{(i)}(w_0)$, $i = 0, \ldots, n$.*

Parametric continuity of order n is denoted C^n. Positional *discontinuity* is denoted C^{-1}; C^0-continuity is defined as positional continuity. Note that a one-segment polynomial curve of arbitrary degree as given by Equation (8.1) is C^∞.

Two curve segments need not have the same derivative vector at their joint in order to have the same tangent line, as illustrated in Figure 8.2. Similarly, they need not be C^2 in order to have the same normal curvature (defined below). A crucial observation here is that derivatives depend on the parameterization while the tangent line and curvature depend on the shape of the spline and are independent of parameterization, i.e. they are *intrinsic*. In order to base the notion of continuity on intrinsic aspects of the curve, we can follow two approaches: take a closer look at the effects of parameterizations (algebraic approach) or take the intrinsic notions like tangent and curvature as a starting point (differential geometric approach).

Since parametric continuity depends on the parameterization, one possibility for an intrinsic notion of continuity is to avoid any dependency on a specific parameterization. This leads to an algebraic definition of continuity called geometric continuity:

DEFINITION 8.3 (GEOMETRIC CONTINUITY) *Curves $P(u)$ and $Q(w)$ are n-th order geometric continuous at u_0 and w_0, if and only if there exists a regular reparameterization $u = u(\tilde{u})$ such that $\tilde{P}(\tilde{u})$ and $Q(w)$ are C^n at $\tilde{P}(\tilde{u}_0)$ and $Q(w_0)$, where $\tilde{P}(\tilde{u}) = P(u(\tilde{u}))$, and $\tilde{u}_0$ is such that $u(\tilde{u}_0) = u_0$.*

Geometric continuity of order n is denoted by G^n or GC^n, and is also called visual continuity. The term visual continuity was first used by [Farin, 82b], and the

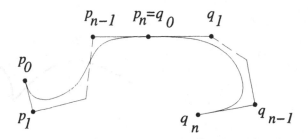

Figure 8.3. G^1-continuity conditions for two Bézier segments.

term geometric continuity by [Barsky and Beatty, 83]. However, the concepts of geometric continuity were already exploited by, for example, [Geise, 62].

The direction of the tangent line is determined by the tangent vector, the normalized derivative vector $T_1(u) = P^{(1)}(u)/\|P^{(1)}(u)\|$, which has unit length. Two curves need not have the same derivative vector in order to have the same tangent vector, as was illustrated in Figure 8.2. The normal curvature vector is defined as

$$T_2(u) = T_1^{(1)}/\kappa_1(u),$$

where $\kappa_1(u)$ is a scalar such that $\|T_2(u)\| = 1$; $\kappa_1(u)$ is called the (scalar) curvature.

First and second order geometric continuity can alternatively be defined as follows:

DEFINITION 8.4 (G^1- AND G^2-CONTINUITY) *Two curves $P(u)$ and $Q(w)$ are G^1-continuous at u_0 and w_0, if and only if their tangent vectors coincide, and G^2-continuous if and only if additionally their normal curvature vectors and scalar curvatures coincide.*

Note that a common tangent line is not sufficient for G^1-continuity, since the tangent vectors must additionally have the same direction. In other words, the two curves must have the same orientation, otherwise they join with a sharp cusp.

8.2.3 Bézier segments

A well known example of blending functions in Equation (8.2) are the Bernstein polynomials, named after [Bernstein, 12]:

(8.3) $$B_i^n(u) = \binom{n}{i} u^i (1-u)^{n-i}, \ 0 \le u \le 1.$$

The resulting curve is the Bézier curve, or segment, of degree n.

The Bézier formulation is often used because of the geometrical significance of the control points, unlike the polynomial coefficients in Equation (8.1). It is

Figure 8.4. Mapping from a parameter domain to a surface.

readily verified that $P(0) = p_0$ and $P(1) = p_n$, so that the curve interpolates the first and last control points. The derivative has the following form:

$$P^{(1)}(u) = n \sum_{i=0}^{n-1} (p_{i+1} - p_i) B_i^{n-1}(u).$$

In particular $P^{(1)}(0) = n(p_1 - p_0)$ and $P^{(1)}(1) = n(p_n - p_{n-1})$, so that the tangent vector at $P(0)$ lies on the line through p_0 and p_1 and the tangent vector at $P(1)$ lies on the line through p_{n-1} and p_n.

Let us consider two Bézier segments P with control points p_i and Q with control points q_i, and a common end point, say $P(1) = Q(0)$, so that $p_n = q_0$. As is just shown, the tangent vector at $P(1)$ has direction $p_n - p_{n-1}$ and the tangent vector at $Q(0)$ has direction $q_1 - q_0$. So, for the two tangent vectors to have the same direction, $p_{n-1}, p_n = q_0$ and q_1 should be collinear, see Figure 8.3.

8.3 Surfaces

8.3.1 Fundamentals

A surface in functional form is a scalar function $f : \mathbb{R}^2 \to \mathbb{R}$. A surface P in parametric form is defined component-wise, each coordinate component being a function of two parameters, i.e. they are bivariate functions. The parameters are allowed to range over some arbitrarily shaped region $D \subseteq \mathbb{R}^2$: $P(s,t) : D \to \mathbb{R}^k$, $k \geq 3$. Thus in particular, the parameter domain need not be rectangular, see Figure 8.4. In its most general form, the parameter domain may have an arbitrary topology with disconnected pieces and holes. Such a cut out (trimmed off) domain corresponds to a so-called *trimmed surface*. Such trimmed surfaces often result from the intersection of two curved surfaces, for example in Constructive Solid Geometry (CSG) modeling, see e.g. [Mäntylä, 88]. The domain over which the surface is defined is then modified, while the coordinate component functions are left unchanged [Casale, 87].

A *polynomial* surface of *total degree* n has the following form:

$$P(s,t) = \sum_{i+j \leq n} p_{i,j} s^i t^j, \quad i,j \in \mathbb{N},$$

with $p_{i,j} \in \mathbb{R}$. A *piecewise* polynomial surface $P(s,t)$ is defined patch by patch. The parameter domain is then partitioned into sub-domains, which can be transformed so as to provide local parameters for each patch. A polynomial *spline* surface is a piecewise polynomial surface.

The i-th partial derivative with respect to s and the j-th with respect to t is denoted $P^{(i,j)}(s,t)$:

$$P^{(i,j)}(s,t) = \frac{\partial^{i+j}P}{\partial s^i \partial t^j}(s,t)$$

Note that $P^{(i,j)}(s,t)$ is a vector. To avoid possible problems with the parameterization of a surface $P(s,t)$, we assume in the following that the first partial derivatives $P^{(1,0)}(s,t)$ and $P^{(0,1)}(s,t)$ exist, and are linear independent. The surface and its parameterization are then said to be *regular*.

The first order partial derivatives are a special case of a *directional* derivative, which is defined as follows:

DEFINITION 8.5 (DIRECTIONAL DERIVATIVE) *The directional derivative at surface point $P(s_0, t_0)$ in the direction $d = (d_s, d_t)$ in the parameter space, is*

$$\nabla_d P(s_0, t_0) = \lim_{h \to 0} \frac{P(s_0 + hd_s, t_0 + hd_t) - P(s_0, t_0)}{h}.$$

The partial derivatives are obtained in the direction of the axes of the parameter space: $\nabla_{(1,0)}P(s_0, t_0) = P^{(1,0)}(s_0, t_0)$, and $\nabla_{(0,1)}P(s_0, t_0) = P^{(0,1)}(s_0, t_0)$.

We are interested in the continuity of spline surfaces $P(s,t)$ and $Q(u,w)$, in particular along a common curve or edge.

8.3.2 Continuity

Analogous to curves, parametric continuity for surfaces is based on the equality of derivatives:

DEFINITION 8.6 (PARAMETRIC CONTINUITY) *Surfaces $P(s,t)$ and $Q(u,w)$ are C^n-continuous at points (s_0, t_0) and (u_0, w_0), if and only if $P^{(i,j)}(s_0, t_0) = Q^{(i,j)}(u_0, w_0)$, $i + j = 0, \ldots, n$.*

The surfaces are C^n along a common curve if they are C^n at each point on that curve. C^{-1} denotes positional discontinuity.

Note that two surfaces need not have the same first order partial derivatives in order to have the same tangent plane. As with curves, the derivatives depend on the parameterization while the tangent plane does not. Moreover, on closed surfaces singularities occur where the derivative of the surface is not defined. For an example see Figure 8.5, which shows a closed piecewise triangular surface P, together with a global parameterization and a partial local parameterization. Corresponding points in the parameter domain and the surface are indicated; the global domain is 'folded' so as to join the D_i to point D on the surface. Taking the derivative at $P(A)$ in the directions

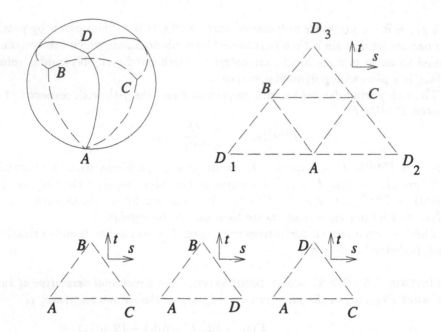

Figure 8.5. Closed surface (top left) with a global parameterization (top right), and a partial local parameterization (bottom row).

$D_1 - A$ and $D_2 - A$ in the global parameterization, it follows from Definition 8.5 that $\nabla_{D_1-A}P(A) \neq \nabla_{D_2-A}P(A)$ (in fact if $D_1 - A = A - D_2$ then $\nabla_{D_1-A}P(A) = -\nabla_{D_2-A}P(A)$, see also [Herron, 85]). By contrast, the derivative at the closed surface requires $\nabla_{D_1-A}P(A) = \nabla_{D_2-A}P(A)$. One might think that a local parameterization solves the problem, but a parameterization of patch (A, B, C) as shown in Figure 8.5, determines $\nabla_{B-A}P(A)$ and $\nabla_{C-A}P(A)$ which imply the parameterization for patch (A, B, D) as shown, which in turn determines $\nabla_{D-A}P(A)$. But the parameterization of (A, B, C) also implies the parameterization of patch (A, C, D) as shown. The resulting $\nabla_{D-A}P(A)$ conflicts with the previous one. Thus, in both the global and the local parameterization the derivative is not properly defined, while the tangent planes may still be coincident at common patch boundaries.

Again we can take an algebraic and a differential geometry approach in order to base the notion of continuity on intrinsic aspects of the surface. Analogous to curves, geometric continuity can be defined algebraically as follows [DeRose and Barsky, 85]:

DEFINITION 8.7 (GEOMETRIC CONTINUITY) *Two surfaces $P(s,t)$ and $Q(u,w)$ are G^n-continuous at (s_0, t_0) and (u_0, w_0) if and only if there exist regular reparameterizations $s = s(\tilde{s}, \tilde{t})$ and $t = t(\tilde{s}, \tilde{t})$, such that $\tilde{P}(\tilde{s}, \tilde{t})$ and $Q(u, w)$ are C^n-continuous at $\tilde{P}(\tilde{s}_0, \tilde{t}_0)$ and $Q(u_0, w_0)$, where $\tilde{P}(\tilde{s}, \tilde{t}) = P(s(\tilde{s}, \tilde{t}), t(\tilde{s}, \tilde{t}))$, and $\tilde{s}_0$ and $\tilde{t}_0$ are such that $s(\tilde{s}_0, \tilde{t}_0) = s_0$ and $t(\tilde{s}_0, \tilde{t}_0) = t_0$.*

Note that the reparameterization may be different at another point $P(s_1, t_1)$, otherwise we would again run into trouble with closed surfaces.

The differential geometry approach to define geometric continuity is based on the surface tangent plane and curvatures. Just as we considered the tangent line of a curve we now consider the tangent plane of a surface. The tangent plane at $P(s_0, t_0)$ is spanned by the derivative vectors $P^{(1,0)}(s_0, t_0)$ and $P^{(0,1)}(s_0, t_0)$. The tangent plane is normal to the surface normal vector

$$(8.4) \qquad N(s_0, t_0) = \frac{P^{(1,0)}(s_0, t_0) \times P^{(0,1)}(s_0, t_0)}{\|P^{(1,0)}(s_0, t_0) \times P^{(0,1)}(s_0, t_0)\|},$$

where '$\times$' denotes the vector or cross product. The tangent planes at $P(s_0, t_0)$ and $Q(u_0, w_0)$ coincide if and only if $P^{(1,0)}(s_0, t_0)$, $P^{(0,1)}(s_0, t_0)$, $Q^{(1,0)}(u_0, w_0)$, and $Q^{(0,1)}(u_0, w_0)$ are coplanar.

However, analogous to curves, a common tangent plane is not sufficient for G^1-continuity, since the surfaces must have the same orientation, i.e. the same unit normal vector. Otherwise they join with a sharp ridge.

THEOREM 8.1 (G^1-CONTINUITY) *Two surfaces are G^1-continuous at a point if and only if their unit normal vectors at that point coincide.*

Second order geometric continuity is based on curvature. For any direction d in the tangent plane at $P(s_0, t_0)$, the plane through d and $N(s_0, t_0)$ intersects $P(s, t)$ in a curve. The normal curvature of this curve is the normal curvature of the surface in the direction of d: $\kappa_d(s_0, t_0)$. Unless $\kappa_d(s_0, t_0)$ is the same in all directions, there are two directions d_1 and d_2 in which $\kappa_d(s_0, t_0)$ takes the maximum and minimum values: the principal curvatures $\kappa_1(s_0, t_0)$ and $\kappa_2(s_0, t_0)$, respectively.

THEOREM 8.2 (G^2-CONTINUITY) *Two surfaces are G^2-continuous at a point if and only if their normal vectors and their principal curvatures at that point coincide.*

8.3.3 Bézier triangles

In the definition of Bézier triangles, the so-called barycentric coordinates are often used as parameters, because of the symmetric form of the resulting definition. Let us first rewrite the univariate Bernstein polynomial $B_i^n(u)$ given in Equation (8.3) into:

$$(8.5) \qquad B_{i,j}^n(t, u) = \frac{n!}{i!j!} t^i u^j, \quad i + j = n, \quad i, j \in \mathbb{N},$$

where t and u are the barycentric coordinates on an arbitrary closed interval $[a, b]$. Any value c in $\mathbb{R}$ can be uniquely expressed in terms of barycentric coordinates with respect to any closed interval: $c = ta + ub$. Note that this is equivalent to $c = a + u(b - a)$. Regarding t and u as weights of a and b, c is the center of gravity, or barycenter. Hence the term barycentric coordinates.

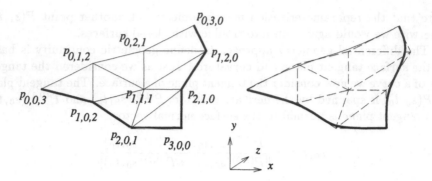

Figure 8.6. Left: Bézier control polyhedron. Right: corresponding cubic Bézier patch.

Analogously, any point D in the plane can be uniquely expressed in barycentric coordinates (t, u, w) with $t + u + w = 1$, relative to three ordered points A, B, and C that are not collinear: $D = tA + uB + wC$. Note that this is equivalent to $D = A + u(B - A) + w(C - A)$. Barycentric coordinates are treated in more detail by [Farin, 86] and [Farin, 90a].

Analogous to the univariate Bernstein polynomials over an interval, the Bernstein polynomials of degree n over a non-degenerate triangle (A, B, C) are defined by

$$B_{i,j,k}^n(t, u, w) = \frac{n!}{i!j!k!} t^i u^j w^k, \quad i + j + k = n, \quad i, j, k \in \mathbb{N},$$

where t, u, and w, with $t + u + w = 1$, are barycentric coordinates with respect to (A, B, C). The Bernstein polynomials form a basis for all polynomials of total degree n over that triangle. That is, every polynomial function $f : (A, B, C) \to \mathbb{R}$ of degree n can be written in the form

$$f(t, u, w) = \sum_{i+j+k=n} p_{i,j,k} B_{i,j,k}^n(t, u, w), \quad p_{i,j,k} \in \mathbb{R}.$$

The polynomial function f describes a surface over the domain triangle (A, B, C). A surface patch in this form is called a Bézier patch, and the scalars $p_{i,j,k}$ are called Bézier ordinates. Such a patch is a function and cannot have an arbitrary shape in 3D. In particular, it cannot be used for interpolation of arbitrary scattered data in 3D, or form a closed surface.

A parametric Bézier triangle in arbitrary dimension is defined componentwise:

$$P(t, u, w) = \sum_{i+j+k=n} p_{i,j,k} B_{i,j,k}^n(t, u, w),$$

where $p_{i,j,k}$ are points in the embedding space, and are called control points, forming an open control polyhedron. See Figure 8.6 for a control polyhedron and the corresponding *cubic* Bézier patch. Note that a parametric patch is defined

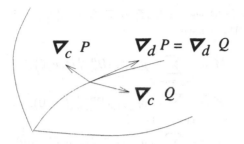

Figure 8.7. Tangent plane continuity is achieved if $\nabla_c P$, $\nabla_d P = \nabla_d Q$, and $\nabla_c Q$ are coplanar.

without explicit reference to a domain triangle. An extensive presentation of triangular Bézier patches is given by [Farin, 86].

The Bézier formulation is often used because of the geometrical significance of the control points. For instance,

$$P(t, u, 0) = \sum_{i+j+k=n} p_{i,j,k} B^n_{i,j,k}(t, u, 0) = \sum_{i+j=n} p_{i,j,0} B^n_{i,j}(t, u),$$

which is a univariate Bézier curve. The patch is thus a curved triangle interpolating $p_{n,0,0}$, $p_{0,n,0}$, and $p_{0,0,n}$, having Bézier curve edges, see Figure 8.6.

The difference between two barycentric coordinates defines a direction in the parameter space. The derivative in the direction $d = (d_1, d_2, d_3)$, $d_1 + d_2 + d_3 = 0$, is given by

(8.6)

$$\nabla_d P(t, u, w) = n \sum_{i+j+k=n-1} (d_1 p_{i+1,j,k} + d_2 p_{i,j+1,k} + d_3 p_{i,j,k+1}) B^{n-1}_{i,jk}(t, u, w).$$

Let us consider two patches P with control points $p_{i,j,k}$ and Q with control points $q_{i,j,k}$, having a common edge. Without loss of generality we may assume that $P(t, u, 0) = Q(t, u, 0)$. The tangent plane of P along $P(t, u, 0)$ is spanned by any two derivative vectors having different directions, for example the directions $d = (0, 1, 0) - (1, 0, 0) = (-1, 1, 0)$ and $c = (0, 0, 1) - (1, 0, 0) = (-1, 0, 1)$. The derivative vector $\nabla_d P$ is actually the tangent vector of $P(t, u, 0)$. Since $P(t, u, 0) = Q(t, u, 0)$, we also have $\nabla_d P = \nabla_d Q$. So, the tangent plane of Q spanned by $\nabla_d Q$ and $\nabla_c Q$ coincides with the tangent plane of P if and only if $\nabla_c Q$, $\nabla_d Q = \nabla_d P$, and $\nabla_c P$ lie in the same plane. That is,

(8.7)

$$[\nabla_d P, \nabla_c P, \nabla_c Q] = 0,$$

where '[ ]' denotes the determinant. Such a situation is depicted in Figure 8.7. Necessary and sufficient conditions on the control points are derived from this constraint by [DeRose, 90]. Note again that Equation 8.7 is weaker than the condition for G^1-continuity, since the orientation of the tangent plane (direction of the normal vector) is not determined here.

Using Equation (8.6) we get $\nabla_d P = \nabla_d Q = n(K - M)$, $\nabla_c P = n(L - M)$, and $\nabla_c Q = n(R - M)$, where

$$
\begin{aligned}
M &= \sum_{i+j=n-1} p_{i+1,j,0} B_{i,j,0}^{n-1}(t, u, 0), \\
K &= \sum_{i+j=n-1} p_{i,j+1,0} B_{i,j,0}^{n-1}(t, u, 0), \\
L &= \sum_{i+j=n-1} p_{i,j,1} B_{i,j,0}^{n-1}(t, u, 0), \text{ and} \\
R &= \sum_{i+j=n-1} q_{i,j,1} B_{i,j,0}^{n-1}(t, u, 0).
\end{aligned}
$$

(8.8)

M, K, L, and R are functions of t, since $u = 1 - t$. We see that $\nabla_d P$, $\nabla_c P$, and $\nabla_c Q$ are all of degree $(n - 1)$.

The requirement that the determinant in Equation (8.7) equals zero amounts to

(8.9) $$(R - M) = \alpha(t)(K - M) + \beta(t)(L - M).$$

Solving $\alpha(t)$ and $\beta(t)$ shows that they are rational polynomial functions having a numerator and denominator of degree $n - 1$ at most. An equivalent formulation for Equation (8.9) is thus the following tangent plane continuity condition:

(8.10)
$$E(t, u)(K - M) + F(t, u)(L - M) + G(t, u)(R - M) = 0, \quad t + u = 1,$$

where E, F, and G are polynomials having at most degree $n - 1$.

The edge that P and Q have in common may be degenerate, that is, of lower degree than the patch itself. Also the two patches may be of different degree. So, the degrees of $\nabla_d P$, $\nabla_c P$, and $\nabla_c Q$ can all be different. The degrees of E, F, and G, and the necessary and sufficient conditions on the control points are derived by [Liu and Hoschek, 89].

8.4   Visual aspects of continuity

Geometric continuity conditions for curves with application to insole shape design are described by [Manning, 74]. G^2-continuity seems to be sufficient for a visually smooth appearance of the curve. A discontinuity in the curvature can be detected by a practised eye, but it seems that higher order discontinuity is not visible.

The visual aspects of surfaces are much more complicated, since the surface contour, its texture, contrast, and illumination, are all involved, see [Koenderink, 90]. Here, I will discuss one aspect of illumination: reflection lines.

Let us see how continuity between surface patches affects the specular reflection of a linear light source. Consider first an illumination model with a point

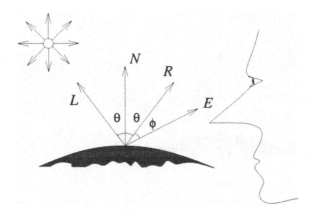

Figure 8.8. Point light source reflection geometry.

light source. For any point on the surface, N is the surface normal, L is the unit vector in the direction of the light, R is the unit reflection vector, and E is the unit vector in the direction of the viewpoint, the eye, see Figure 8.8. A mirror reflects light from direction L only in the direction R. Glossy surfaces exhibit specular reflection by scattering light anisotropically. The empirical model for the specular reflection of glossy surfaces described by [Phong, 75] assumes that maximum specular reflection occurs when the angle ϕ between R and the eye vector E is zero, and falls off sharply as ϕ increases.

Now, given a linear light source $LL(t)$, the shape of the reflection curve $LL^*(t)$ depends on the shape of the surface. If the surface is G^1-continuous, the tangent vectors of $LL^*(t)$ at both sides of the patch boundary lie in the surface tangent plane, but need not be collinear, see Figure 8.9. So, G^1 surfaces generally show a reflection line that is not G^1-continuous.

Conversely, let us now assume that $LL^*(t)$ is G^1-continuous, as well as the surface itself. Let V be the view point (position of the eye) such that the angle

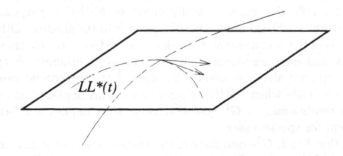

Figure 8.9. G^1 surface with a reflection curve $LL^*(t)$ that is *not* G^1.

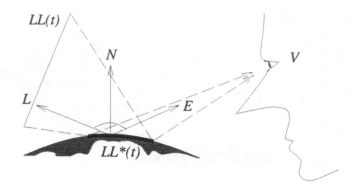

Figure 8.10. Linear light source reflection geometry with $\phi = 0$.

between E and R is zero, then

$$E(t) = \frac{V - LL^*(t)}{\|V - LL^*(t)\|}$$

and

$$L(t) = \frac{LL(t) - LL^*(t)}{\|LL(t) - LL^*(t)\|}.$$

A necessary and sufficient condition for a point $LL^*(t)$ to be a reflection point
is (see Figure 8.10):

(8.11) $N(t) \cdot (E(t) - L(t)) = 0,$

where $N(t)$ is the *surface* normal along the reflection curve, and '$\cdot$' the dot or
scalar product of two vectors. Differentiation of Equation (8.11) with respect to
t gives

(8.12) $N^{(1)}(t) \cdot (E(t) - L(t)) + N(t) \cdot (E^{(1)}(t) - L^{(1)}(t)) = 0.$

This holds if both terms of the sum are zero, so that either $N^{(1)}(t) = 0$, which
means that the reflection curve is lically linear, or $N^{(1)}(t)$ is perpendicular to
$(E(t) - L(t))$, which whould imply that the point is in the shadow. Otherwise, if
not both terms of the sum are zero, then since the surface is G^1-continuous, $N(t)$
is continuous, and since we assume that $LL^*(t)$ is G^1-continuous, $E(t)$, $E^{(1)}(t)$,
$L(t)$, and $L^{(1)}(t)$ are also continuous. So, apart from the special cases, Equation (8.12) only holds when $N^{(1)}(t)$ is continuous. This implies that the surface
curvature is continuous. A G^1 reflection curve thus implies a G^2-continuous
surface, except for special cases.

On the other hand, G^2-continuity of the surface does not imply satisfaction
of Equation (8.12), and so Equation (8.11) need not hold. Therefore, a G^2-
continuous surface need not imply a G^1 reflection curve.

If the angle ϕ between E and R is larger than zero, the same effect is visible,
except for a sharp falloff in the reflected light intensity.

8.5   Concluding remarks

This chapter has introduced representations for curves and surfaces, in particular the Bézier formulation, and the notions parametric and geometric continuity. I have shown that derivatives at closed surfaces exhibit singularities for piecewise parameterization in addition to the case for global parameterization, so that parametric continuity is not even properly defined.

I have related geometric continuity to illumination, and have shown in particular that a G^1 surface need not give a G^1 reflection curve, and on the other hand, a G^1 reflection curve implies that the surface is G^2. So, the surface can be visually inspected without explicitly computing curves of constant curvature along the surface, or other surface features [Higashi et al., 90]. Incorporating a linear light source into the illumination model gives surface continuity information for free. However, algorithms interpolating the surface normal vector, like Phong shading, may introduce artifacts [Foley et al., 90].

Perception of (dis)continuity of a surface depends not only on the surface itself but also on the illumination, in a computer model as well as in the physical world. The term 'visual continuity' for what has been defined as G^n-continuity is therefore inappropriate.

9

G^1 boundary construction

This chapter is concerned with the construction of a G^1-continuous object boundary. For 2D, a straightforward construction of a closed piecewise cubic Bézier curve passing through given vertices with prescribed tangent vectors is presented. For 3D, an analysis of the total degree required to solve several interpolation problems using polynomial patches is given. The attention is then focused on the construction of a closed piecewise triangular cubic Bézier surface, that interpolates given vertices with prescribed normal vectors. In order to get sufficient degrees of freedom to define the control points, a triangle three-split, a two-split and a six-split scheme are developed. The split into six sub-triangles results in a surface that is G^1-continuous as well as visually pleasing.

9.1 Introduction

We can distinguish several scattered data interpolation problems, depending on the input data and the continuity requirements. In each case our purpose is to construct a smooth closed boundary through the vertices.

In 2D, the input must at least consist of the topology of a set of vertices along the boundary, i.e. an ordering. This is equivalent to a closed polygon, for example obtained by a reconstruction (Chapter 5) or polygonal approximation algorithm (Chapter 7). Additional data at the vertices can be the tangent line or derivative vectors. Usual continuity requirements are G^1- or G^2-continuity everywhere at the curve.

In 3D, the input must at least consist of the topology of a set of vertices along the boundary, i.e. a triangulation of the vertices. This is equivalent to a closed polyhedron of triangular facets, for example obtained by a reconstruction (Chapter 5) or polyhedral approximation algorithm (Chapter 7). Additional data at the vertices can be the (unit) surface normal, tangent vectors (direction of derivatives) of the patch edges, or derivative vectors (tangent and magnitude) of the edges. Data along the edges are for example surface normals, cross edge derivatives and curvatures. Usual continuity requirements are G^1- or G^2-continuity everywhere at the surface.

In this chapter we will consider the interpolation of the vertices of the polyhedron by triangular patches. I shall refer to the interpolation of vertex positions as the P-interpolation problem, to position and surface normal at the vertices as the PN-interpolation problem, to position and patch edge tangent vector as the PT-interpolation problem, and to position and patch edge derivative vector as the PD-interpolation problem. Note that PN-interpolation not only requires a common tangent plane of all patches incident to a vertex, but also the same orientation of that tangent plane. A surface that satisfies the PN-interpolation requirements is G^1-continuous at the vertices. In order to be G^1-continuous at every point on the surface, the patches should be G^1 themselves, and should join G^1-continuously at the edges. Note further that PT-interpolation does *not* denote the interpolation of the tangent *plane*, but rather the tangent vectors along the patch edges, which is more restrictive.

The rest of this chapter is organized as follows. Section 9.2 gives a simple and obvious way to construct a G^1-continuous boundary curve. Section 9.3 presents an analysis of the required polynomial degree for the various interpolation problems in 3D, not found in the literature. Section 9.5 gives an overview of existing local methods for the PN-interpolation problem. Section 9.6 introduces a new solution that is cubic and based on the splitting of a triangle into three sub-triangles, and Section 9.7 describes how the triangle splitting can be made adaptive, that is, dependent on the geometry of the triangulation. Section 9.8 presents a scheme that splits a triangle into six sub-triangles, and Section 9.9 gives some concluding remarks.

9.2   A G^1 boundary curve

Consider a sequence of vertices $v_0, \ldots, v_{N_v-1}$ in the plane that has to be interpolated. A polygon through the vertices is a linear interpolation with only C^0-continuity. Our purpose is to construct a closed G^1 curve through the vertices. This is done in the following way. First we estimate the tangent vector at each vertex, then we construct a G^1 curve that interpolates vertices and tangent vectors.

Let us denote the tangent vector at v_i with T_i. This tangent line can be estimated by weighting the vectors $(v_i - v_{i-1})/\|v_i - v_{i-1}\|$ and $(v_{i+1} - v_i)/\|v_{i+1} - v_i\|$ and normalize the sum:

1. Weight by length: take $(v_i - v_{i-1}) + (v_{i+1} - v_i)$, giving $v_{i+1} - v_{i-1}$, and normalize to unit length. The reasoning behind this method is that the larger

of the segments $v_{i-1}v_i$ and v_iv_{i+1} corresponds to a larger part of the curve, and should affect the tangent vector most.

2. Weight uniformly: take $(v_i - v_{i-1})/\|v_i - v_{i-1}\| + (v_{i+1} - v_i)/\|v_{i+1} - v_i\|$, and normalize to unit length.

3. Weight by inverse length: take $(v_i - v_{i-1})/\|v_i - v_{i-1}\|^2 + (v_{i+1} - v_i)/\|v_{i+1} - v_i\|^2$, and normalize to unit length. The idea of this method is that a close neighboring vertex knows more about the local curve tangent and should have a larger weight than the far vertex.

All these methods only take into account the neighboring vertices v_{i-1} and v_{i+1}. Other methods could be applied that use more vertices and for example estimate the tangent line by a least-squares fit. The analogue of method 1 for surface normal estimation has been reported to work well, see Section 9.4; therefore method 1 is used here.

In order to get a closed G^1 curve we construct a Bézier segment between each pair of consecutive vertices. Let us consider the degree n Bézier segments P between v_{i-1} and v_i, and Q between v_i and v_{i+1} (in this subsection the indices are considered modulo N_v). In order to interpolate the vertices, we must set $p_0 = v_{i-1}$, $p_n = q_0 = v_i$, and $q_n = v_{i+1}$. For the derivative vectors $P^{(1)}(1)$ and $Q^{(1)}(0)$ to be collinear, p_{n-1} and q_1 must lie on the line through T_i, denoted by $Tline_i$. If $Tline_{i-1}$ intersects $Tline_i$, and $Tline_i$ intersects $Tline_{i+1}$, we can set $n = 2$, p_1 to $Tline_{i-1} \cap Tline_i$, and q_1 to $Tline_i \cap Tline_{i+1}$. The resulting Bézier segments are then completely defined, and have collinear tangent vectors. All other segments are defined in the same way, so that a closed quadratic curve is constructed. However, with only tangent *line* continuity the tangent *vectors* can still have opposite directions, which gives sharp cusps, i.e. not G^1-continuity.

In order to ensure G^1-continuity we need more degrees of freedom. This is obtained by using one more control point, so that we get cubic Bézier segments. Now we must set $p_0 = v_{i-1}$, $p_3 = q_0 = v_i$, and $q_3 = v_{i+1}$, to interpolate the vertices. The control points p_1 and p_2 are determined by the following heuristic method, which gives good results in many practical cases. Vertex p_3 is orthogonally projected onto $Tline_{i-1}$, giving p_3^*. Control point p_1 is then set to $p_0 + (p_3^* - p_0)/3$. Analogously, $p_2 = p_3 + (p_0^* - p_3)/3$, where p_0^* is the orthogonal projection of p_0 onto $Tline_i$. All other segments are defined in the same way, so that a closed cubic curve is constructed. The curve is G^1-continuous at v_i if $P^{(1)}(1)$ and $Q^{(1)}(0)$ have the same direction, which is the case when p_3^* and q_0^* lie at opposite sides of v_i on $Tline_i$. This condition is satisfied for many sets of vertices whose successive tangent lines do not change direction wildly. In other cases, the tangent line at a vertex should be changed in order for the projection method above to work properly.

Figure 9.1 shows an example of this algorithm to construct a G^1-continuous curve, using method 1 to estimate the tangent line.

Cubic curves are necessary and sufficient to achieve even G^2-continuity, see [Farin, 82b] and [Böhm, 85].

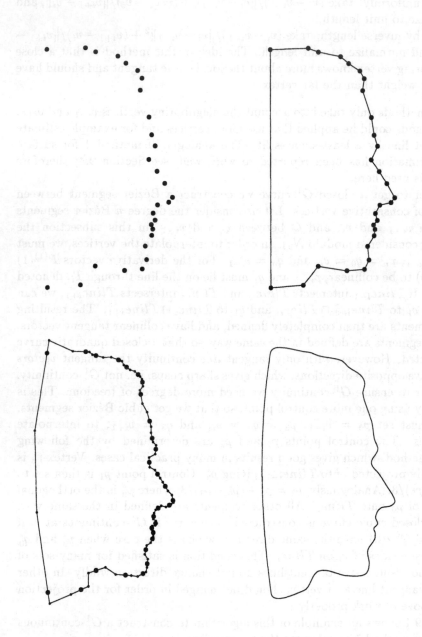

Figure 9.1. Vertices and linear interpolation (above), and the constructed Bézier control polygon and G^1 curve (below).

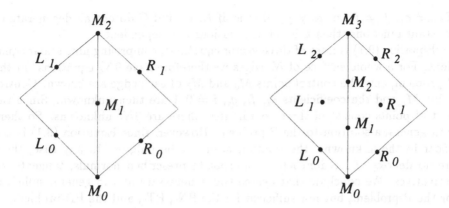

Figure 9.2. Control points involved for G^1 connection of two quadratic (left) and cubic (right) patches.

9.3 Analysis of surface degree

A polyhedron is a linear interpolation of the vertices with only C^0-continuity. One may wonder what polynomial degree is necessary for the various G^1 interpolation problems PN, PT, and PD. It has been shown by [Piper, 87] that degree four is sufficient for the PD problem. He has further shown by means of a counterexample, but not by analysis, that degree three is not always sufficient. In this section I present an analysis of the required polynomial degree for the various interpolation problems in 3D.

Let us consider the G^1-continuity conditions for two Bézier patches P and Q that have a common edge, say $P(u, v, 0) = Q(u, v, 0)$. The control points involved in the G^1 connection of two patches are shown in Figure 9.2. To simplify notation, we denote $p_{i,j,0} = q_{i,j,0}$ with M_i (middle column of control points), $p_{i,j,1}$ by L_i (left), and $q_{i,j,1}$ by R_i (right).

Quadratic case. For quadratic patches M, K, L, and R (given by Equation (8.8)) are: $M = tM_0 + uM_1$, $K = tM_1 + uM_2$, $L = tL_0 + uL_1$, and $R = tR_0 + uR_1$. The functions E, F, and G are at most linear: $E(t, u) = e_0 t + e_1 u$, $F(t, u) = f_0 t + f_1 u$, and $G(t, u) = g_0 t + g_1 u$. Since $u = 1 - t$, we see that $E(t, u) = t(e_0 - e_1) + e_1$ reduces to a constant if $e_0 = e_1$, and likewise for F and G.

Substitution of M, K, L, R, E, F, and G into the tangent plane continuity condition Equation (8.10) yields $t^2 C_0 + tuC_1 + u^2 C_2 = 0$, with coefficients C_i as given below. Since this equation must hold for all $t + u = 1$, C_0, C_1, and C_2 should all be zero:

$$C_0 = e_0(M_1 - M_0) + f_0(L_0 - M_0) + g_0(R_0 - M_0) = 0,$$

(9.1)
$$C_1 = e_1(M_1 - M_0) + e_0(M_2 - M_1) + f_1(L_0 - M_0) + f_0(L_1 - M_1) +$$
$$g_1(R_0 - M_0) + g_0(R_1 - M_1) = 0,$$
$$C_2 = e_1(M_2 - M_1) + f_1(L_1 - M_1) + g_1(R_1 - M_1) = 0.$$

If $e_0 \neq e_1$, $f_0 \neq f_1$, or $g_0 \neq g_1$, that is, if E, F, and G do not all degenerate to constant functions, then this set of equations is independent.

Equation (9.1) is a set of three vector equations, comprising nine scalar equations. For a whole surface of N_e edges we therefore have $9N_e$ equations. In the P-problem, only the control points M_0 and M_2 of each edge are known. Control point M_1 and the coefficients e_i, f_i, g_i, $i = 0, 1$ are then unknown. Since the control points consist of three coordinates, there are $9N_e$ unknowns. So there is in general a solution to the P-problem. However, since Equation (9.1) is not linear in the unknowns, the solution need not be unique. In any case, there are no degrees of freedom left to interpolate prescribed normals, tangents, or derivatives. We conclude that degree two is necessary and in general sufficient for the P-problem, but not sufficient for the PN-, PT-, and the PD-problem.

Cubic case. For cubic patches M, K, L, and R (given by Equation (8.8)) are:
$M = t^2 M_0 + 2tu M_1 + u^2 M_2$, $K = t^2 M_1 + 2tu M_2 + u^2 M_3$, $L = t^2 L_0 + 2tu L_1 + u^2 L_2$, and $R = t^2 R_0 + 2tu R_1 + u^2 R_2$. The functions E, F, and G are at most quadratic: $E(t, u) = e_0 t^2 + e_1 tu + e_2 u^2$, etc. Since $u = 1 - t$, we see that $E(t, u) = t^2(e_0 - e_1 + e_2) + t(e_1 - 2e_2) + e_2$ reduces to a linear function if $e_0 - e_1 + e_2 = 0$, and to a constant if additionally $e_1 - 2e_2 = 0$.

Substitution of M, K, L, R, E, F, and G into the tangent plane continuity condition Equation (8.10) now gives $t^4 C_0 + t^3 u C_1 + t^2 u^2 C_2 + tu^3 C_3 + u^4 C_4 = 0$, with

(9.2a)
$$C_0 = e_0(M_1 - M_0) + f_0(L_0 - M_0) + g_0(R_0 - M_0) = 0,$$

(9.2b)
$$C_1 = e_1(M_1 - M_0) + 2e_0(M_2 - M_1) + f_1(L_0 - M_0) +$$
$$2f_0(L_1 - M_1) + g_1(R_0 - M_0) + 2g_0(R_1 - M_1) = 0,$$

(9.2c)
$$C_2 = e_2(M_1 - M_0) + 2e_1(M_2 - M_1) + e_0(M_3 - M_2) +$$
$$f_2(L_0 - M_0) + 2f_1(L_1 - M_1) + f_0(L_2 - M_2) +$$
$$g_2(R_0 - M_0) + 2g_1(R_1 - M_1) + g_0(R_2 - M_2) = 0,$$

(9.2d)
$$C_3 = 2e_2(M_2 - M_1) + e_1(M_3 - M_2) + 2f_2(L_1 - M_1) +$$
$$f_1(L_2 - M_2) + 2g_2(R_1 - M_1) + g_1(R_2 - M_2) = 0,$$

(9.2e) $$C_4 = e_2(M_3 - M_2) + f_2(L_2 - M_2) + g_2(R_2 - M_2) = 0.$$

If $e_0 - e_1 + e_2 \neq 0$, $f_0 - f_1 + f_2 \neq 0$, and $g_0 - g_1 + g_2 \neq 0$, i.e. if E, F, and G do not all degenerate to linear functions, then this set of equations is independent.

Equation (9.2) is a set of five vector equations, or fifteen scalar equations. For a whole surface of N_e edges and N_t triangles we therefore have $15N_e$ equations. Considering first the P-problem, only the control points M_0 and M_3 of each edge are known. Control points M_1 and M_2 and the coefficients e_i, f_i, g_i, $i = 0, 1, 2$ are unknown, and for each triangle the control point L_1 is also unknown. So, there are $15N_e + 3N_t$ unknowns. However, for a closed triangulation $3N_t = 2N_e$.

This results in a total of $15N_e$ equations in $17N_e$ unknowns, so that in general there is more than one solution.

The PN-problem prescribes that the control points M_1 and M_2 lie in given tangent planes at the vertices M_0 and M_3, respectively (apart from the tangent plane orientation). This results in $2N_e$ additional equations, giving a total of $17N_e$ equations in $17N_e$ unknowns. So in general there is a solution to the PN-problem, and if E, F, and G do not degenerate to linear functions, this solution is unique. In that case, however, there are no degrees of freedom left to interpolate prescribed tangents or derivatives. We conclude that degree three is generally necessary and sufficient for the PN-problem, but not sufficient for the PT- and PD-interpolation problem.

[Piper, 87] has shown that degree four is necessary and sufficient for the PD-interpolation problem of *two* patches, but analogous to the previous analysis it is easily verified that it also applies to a whole surface. Consequently, also for the PT-problem degree four is sufficient, and necessary as well, as shown above.

The results of this section are summarized in the following table, giving the necessary and sufficient polynomial degrees for the considered interpolation problems:

	2	3	4
P	+		
PN		+	
PT			+
PD			+

9.4 Surface normal estimation

In the rest of this chapter only the PN-interpolation problem is considered. First, the surface normals at the vertices must be estimated, then we construct a G^1 surface that interpolates the vertices and the surface normals. Analogously to the tangent vector of a curve, the normal vector at a vertex v_i can be estimated by weighting the unit normals of all the incident triangles, and normalize the sum:

1. Weight by area. For each triangle, take the cross-product of two different vectors between its vertices. This vector is normal to the triangle, and its magnitude is twice the triangle area. Sum the vectors, and normalize to unit length. The idea of this method is that the larger triangles correspond to a larger part of the surface, and should affect the orientation of the tangent plane most.

2. Weight uniformly. Divide the cross-products by twice the area of the triangle so as to give unit normals. Then normalize the sum.

3. Weight by inverse area. Divide each unit normal by the area of the triangle, then normalize the sum. The reasoning behind this method is that a close neighboring vertex knows more about the local surface normal and should have a larger weight than far away vertices.

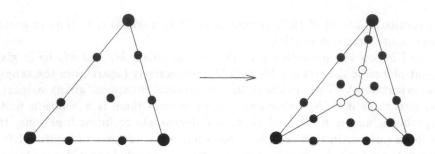

Figure 9.3. Schematic representation of a three-split of a cubic Bézier triangle.

All these methods only take into account the vertices adjacent to v_i. Other methods could be applied that use more vertices, and for example estimate the tangent plane by a least-squares fit. It has been reported that preliminary results indicate that estimation method 1 is the most accurate method [Sloan, 91], and is the one used here.

9.5   Local schemes

Although degree three is sufficient for a solution to the PN problem to exist, it is a global solution, resulting from a large set of equations involving all the control points. Local schemes are preferred because they are simpler, computationally cheaper, and allow local changes of vertex position and normal.

In order to achieve local solutions to the PN-interpolation problem we need more degrees of freedom to choose the control points so as to obtain G^1-continuity. There are three well known strategies to get more degrees of freedom: blending of patches, raising the degree of the polynomial patch from three to four, and patch subdivision. Blended patches are composed of a sum of patches giving an interpolating result, and a correction term to make the patch G^1. However, the correction term is a rational polynomial. Blending methods are applied by for example [Herron, 85], [Nielson, 87], and [Hagen and Pottmann, 88].

Raising of the polynomial degree of the patch is performed by [Farin, 83], [Piper, 87], [Jensen, 87], [Pfluger and Neamtu, 91], and [Schmitt et al., 91]. Some of these methods introduce a degeneracy: [Farin, 83] and [Jensen, 87] let the edge of the patch be of actual degree three (as a result the connection to cubic rectangular patches is straightforward, see [Farin, 82a]), while [Pfluger and Neamtu, 91] and [Schmitt et al., 91] contract some of the interior control points into one.

General patch subdivision splits a patch into several patches that together have the same shape as the original one. Subdivision algorithms for Bézier triangles are given by [Goldman, 83]. The so-called Clough–Tocher triangle splitting scheme (named after [Clough and Tocher, 65]) subdivides a triangular patch $P(t, u, w)$ at the surface point $P(\frac{1}{3}, \frac{1}{3}, \frac{1}{3})$ into three new triangles. The parent triangle is referred to as the macro triangle, the three new ones as micro tri-

angles. See Figure 9.3 for a schematic representation of a three-split of a cubic
Bézier triangle.

The Clough–Tocher split has been used for cubic functional surfaces in finite-
element analysis, see [Strang and Fix, 73], and later for quartic parametric surfaces in scattered data interpolation by [Farin, 83]. The triangle split has two
effects: the number of control points is increased, and the interpolation constraints apply to only one edge of the micro triangle, the one that coincides
with the macro triangle edge. At the interior edges no interpolation data is
prescribed, and only G^1-continuity is required. Consequently, the control points
that are not incident to a macro triangle edge in the control polyhedron, can
be moved freely without affecting interpolation and continuity along the macro
triangle edges.

This three-split scheme is applied by [Farin, 83], [Piper, 87], and [Jensen, 87]
to quartic Bézier patches in order to achieve a local solution to the PN-
interpolation problem. In the following section I will show that a *cubic* three-split
scheme generally provides enough degrees of freedom to solve the PN-problem
locally.

9.6   A cubic three-split scheme

Let a set of vertices be given, as well as their topology (a triangulation) and
surface normals at the vertices. The notation used and the control point lay-out
and naming are illustrated in Figure 9.4. Let us consider the macro triangle
edge between M_0 and M_3, and let E, F, and G be linear: $E(t, u) = e_0 t + e_1 u$,
$F(t, u) = f_0 t + f_1 u$, $G(t, u) = g_0 t + g_1 u$. The tangent plane continuity condition
Equation (8.10) then becomes $t^3 C_0 + t^2 u C_1 + tu^2 C_2 + u^3 C_3 = 0$, with

(9.3a) $C_0 = e_0(M_1 - M_0) + f_0(L_0 - M_0) + g_0(R_0 - M_0) = 0,$

(9.3b) $C_1 = e_1(M_1 - M_0) + 2e_0(M_2 - M_1) + f_1(L_0 - M_0) + 2f_0(L_1 - M_1) +$
$\qquad g_1(R_0 - M_0) + 2g_0(R_1 - M_1) = 0,$

(9.3c) $C_2 = 2e_1(M_2 - M_1) + e_0(M_3 - M_2) + 2f_1(L_1 - M_1) + f_0(L_2 - M_2) +$
$\qquad 2g_1(R_1 - M_1) + g_0(R_2 - M_2) = 0,$

(9.3d) $C_3 = e_1(M_3 - M_2) + f_1(L_2 - M_2) + g_1(R_2 - M_2) = 0.$

The algorithm consisting of the following six steps splits each macro triangle
into three cubic micro triangles, and sets the Bézier control points so as to satisfy
the tangent plane continuity condition along all the generated Bézier patch edges,
that is, all the micro triangle edges.

Step 1.   First, we construct cubic macro triangles P, interpolating the given
vertices. To this end, $p_{3,0,0}$, $p_{0,3,0}$, and $p_{0,0,3}$ are set to the vertices of a given
triangle.

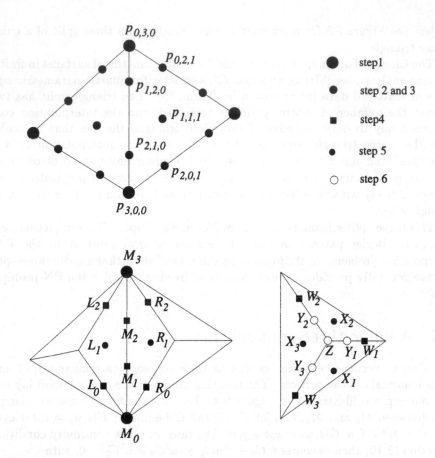

Figure 9.4. Control points used in the three-split scheme.

Step 2. Control point $p_{2,1,0}$ is set as follows: $p_{0,3,0}$ is projected onto the tangent plane at $p_{3,0,0}$ giving a point $p_{0,3,0}^*$, then

$$p_{2,1,0} = p_{3,0,0} + (p_{0,3,0}^* - p_{3,0,0})/3.$$

Control points $p_{1,2,0}$, $p_{0,2,1}$, $p_{0,1,2}$, $p_{2,0,1}$, and $p_{1,0,2}$ are set in a symmetrical way.

Step 3. Following [Farin, 83], $p_{1,1,1}$ is set to as follows:

$$p_{1,1,1} = (p_{2,1,0} + p_{1,2,0} + p_{2,0,1} + p_{1,0,2} + p_{0,2,1} + p_{0,1,2})/4 - (p_{3,0,0} + p_{0,3,0} + p_{0,0,3})/6.$$

We now have a surface interpolating the vertices and normals, but it is not yet tangent plane continuous.

Step 4. Each macro triangle is split at $P(\frac{1}{3}, \frac{1}{3}, \frac{1}{3})$ into three micro triangles by Bézier patch subdivision. For each macro triangle edge the control points M_i, $i = 0, 1, 2, 3$ are computed and remain fixed. R_0 is set to $\frac{1}{3}(p_{3,0,0} + p_{2,1,0} + p_{2,0,1})$, and L_0, L_2, and R_2 are set analogously.

Step 5. Control points L_1 and R_1 have to be set such that the tangent plane continuity condition Equation (8.10) is satisfied. Since M_i, $i = 0, 1, 2, 3$, and L_i, R_i $i = 0, 2$ are known by now, Equations (9.3a) and (9.3d) determine the coefficients e_i, f_i, and g_i, $i = 0, 1$, up to a constant factor, so that we can arbitrarily set $e_0 = e_1 = 1$. The unknowns L_1 and R_1 are constrained by (9.3b) and (9.3c). Rewriting gives:

$$
\begin{aligned}
2f_0 L_1 + 2g_0 R_1 = {} & 2f_0 M_1 + 2g_0 M_1 - e_1(M_1 - M_0) - 2e_0(M_2 - M_1) \\
& - f_1(L_0 - M_0) - g_1(R_0 - M_0), \\
2f_1 L_1 + 2g_1 R_1 = {} & 2f_1 M_1 + 2g_1 M_1 - 2e_1(M_2 - M_1) - e_0(M_3 - M_2) \\
& - f_0(L_2 - M_2) - g_0(R_2 - M_2).
\end{aligned}
\tag{9.4}
$$

If this pair of equations is independent, it uniquely determines L_1 and R_1. Tangent plane continuity is then achieved across the macro triangle edges.

Step 6. In order to get tangent plane continuity across the edges of the micro triangles we set the control points Y_1, Y_2, Y_3, and Z from Figure 9.4 as follows (see [Farin, 83]):

$$
\begin{aligned}
Y_1 &= (W_1 + X_1 + X_2)/3, \\
Y_2 &= (W_2 + X_2 + X_3)/3, \\
Y_3 &= (W_3 + X_3 + X_1)/3, \\
Z &= (Y_1 + Y_2 + Y_3)/3.
\end{aligned}
$$

Correctness of this algorithm is proved by the following theorem.

THEOREM 9.1 *Steps 1 to 6 above construct an overall tangent plane continuous surface.*

Proof. By construction (Step 5), the macro triangle edges satisfy Equation (9.3), and are thus tangent plane continuous. In Step 4, W_3 $(= R_0)$ is set to $\frac{1}{3}(p_{3,0,0} + p_{2,1,0} + p_{2,0,1})$, and in Step 6, Y_3 is set to $\frac{1}{3}(W_3 + X_1 + X_3)$, and Z to $\frac{1}{3}(Y_0 + Y_1 + Y_2)$. This makes the patch tangent plane continuous along the micro triangle edge between $p_{3,0,0}$ and Z, since Equation (8.10), when applied to this edge, is satisfied for constant values of $E(t, u)$, $F(t, u)$, and $G(t, u)$. The other two edges are treated in a symmetrical way in Steps 4 and 6. Because the same constants are used for the other two edges, there is no conflict in setting Z.

The surface is thus tangent plane continuous along all triangle edges. Since cubic Bézier triangles are internally C^2-continuous, the whole surface is tangent plane continuous. $\square$

Essentially the same scheme of splitting triangles into three sub-triangles has been described by [Cottin and Damme, 90], but is independently derived here.

Analogous to the construction of cubic G^1 curves, in order to achieve a continuously changing unit normal vector (G^1-continuity), the orientation of the tangent plane must be properly defined at every point on the common edge of

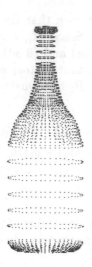
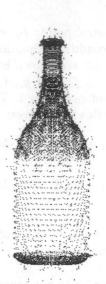

Figure 9.5. Result of the three-split scheme. Left: macro triangle Bézier control points before splitting. Right: final control points of all micro triangles.

two adjacent triangles, in addition to the tangent plane continuity condition. Only when both conditions are met, the surface is G^1-continuous. The tangent plane orientation will be properly defined if f_0 and g_0 in Equation (9.3a) (and also f_1 and g_1 in (9.3d)) have opposite signs. This condition is satisfied for many sets of vertices whose neighboring normal vectors directions do not change wildly. In other cases, the normal vector at a vertex should be altered in order for the projection in Step 2 to result in opposite signs of f_0 and g_0. A detailed analysis on this is presented by [Cottin and Damme, 90].

We see that a three-split gives sufficient degrees of freedom to construct a tangent plane continuous surface of cubic patches. The tangent plane continuity along the macro triangle edges is enforced by setting the control points L_1 and R_1 so as to satisfy Equation (9.4). It turns out, however, that the resulting position of L_1 and R_1 is often far away from their previous position. Apparently L_1 and R_1 often have enough room to satisfy Equation (9.4) only when the plane through M_1, M_2, L_1, and R_1 is very tilted (with respect to its previous orientation). Although the surface tangent plane is then continuous along the common edge, the surface oscillates wildly in such cases. Even if the surface tangent plane additionally has a continuous orientation, i.e. the surface is G^1-continuous, it does not give a visually smooth impression.

This is illustrated in Figure 9.5, showing the result of the algorithm applied to the triangular polyhedral surface of the bottle in Figure 5.10 (left). The image at the left shows the Bézier control points of the macro triangles before splitting.

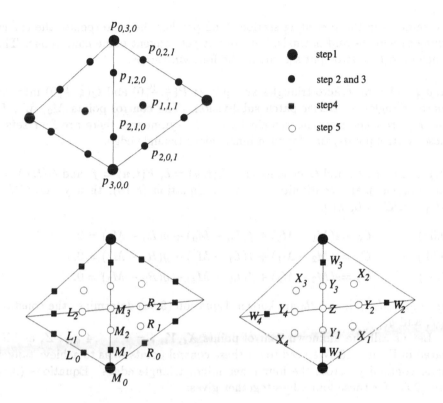

Figure 9.6. Control points used in the two-split scheme.

The right image depicts the final control points, i.e. when the resulting surface is tangent plane continuous.

The reason for this unsatisfactory result is that Equation (9.4) imposes too strict constraints on L_1 and R_1. This could be avoided if there were additional degrees of freedom that could be used to select a solution that is optimal in some sense. Such a solution is presented in Section 9.8.

9.7 Towards an adaptive splitting scheme

Another disadvantage of the three-split is that very thin triangles can emerge. This may cause numerically instable computations.

We can subdivide a thin macro triangle into *two* micro triangles instead of three, by creating a new edge from one vertex to a point on the opposite edge. The neighboring macro triangle at the side of the split edge should also be subdivided. Figure 9.6 shows a schematic representation of a two-split of two cubic Bézier triangles.

Let us consider two patches P and Q such that $P(t,u,0) = Q(t,u,0)$, which are each split into two micro triangles, while the four neighboring macro triangles incident to the other edges are split into three micro triangles. Steps 1 to 3 are the same as in the preceding section. The patches then interpolate the (macro triangle) vertices and normals, but are not yet tangent plane continuous. That is achieved by a two-split scheme in the following steps.

Step 4. The two macro triangles are split at $P(\frac{1}{2}, \frac{1}{2}, 0)$ and $Q(\frac{1}{2}, \frac{1}{2}, 0)$ into four micro triangles by Bézier patch subdivision. The control points M_0, M_1, L_0, and R_0 are computed and remain fixed. By symmetry, there are four sets of these control points: one for each inner micro triangle edge.

Step 5. Let E, F, and G be constant: $E(t, u) = e$, $F(t, u) = f$, and $G(t, u) = g$. The tangent plane continuity condition, Equation (8.10), then yields $t^2 C_0 + tu C_1 + u^2 C_2 = 0$, with

(9.5a) $C_0 = e(M_1 - M_0) + f(L_0 - M_0) + g(R_0 - M_0) = 0,$
(9.5b) $C_1 = e(M_2 - M_1) + f(L_1 - M_1) + g(R_1 - M_1) = 0,$
(9.5c) $C_2 = e(M_3 - M_2) + f(L_2 - M_2) + g(R_2 - M_2) = 0.$

Since M_0, M_1, L_0, and R_0 are known, Equation (9.5a) determines the constants e, f, and g.

Let us call the unknown control points X_i, Y_i, $i = 1, \ldots, 4$ and Z, as illustrated in Figure 9.6. We need to set these control points so as to achieve tangent plane continuity across the four inner micro triangle edges. Equations (9.5b) and (9.5c) for these four edges together gives:

(9.6a) $e_i(Y_i - W_i) + f_i(X_{i-1} - W_i) + g_i(X_i - W_i) = 0,$
(9.6b) $e_i(Z - Y_i) + f_i(Y_{i-1} - Y_i) + g_i(Y_{i+1} - Y_i) = 0,$

for $i = 1, \ldots, 4$, and $i-1$ and $i+1$ taken modulo 4. This results in eight equations in nine unknowns. This set of equations in general has a solution. There are even enough degrees of freedom to choose a 'best' solution, by optimizing a suitable object function. For example, minimizing

$$\sum_{i=1}^{4} \|X_i - X_i^{old}\| + \sum_{i=1}^{4} \|Y_i - Y_i^{old}\| + \|Z - Z^{old}\|,$$

where Z^{old} is the value of Z just before Step 5, and likewise for X_i and Y_i, gives a solution to Equation (9.6) that is close to the old configuration of control points. This is considered good, since the old values were chosen in a sensible way.

The four micro triangles are now tangent plane continuously connected to each other, but must also be tangent plane continuously connected to the four neighboring patches at the four outer edges of the two macro triangles. However, the inner control points of the micro triangles have already been fixed in the previous step. Therefore Step 5 of the three-split scheme must be adapted when used in combination with the two-split scheme. Let us consider one such outer edge, and switch again to the notation of Figure 9.2. We have seen that choosing E, F, and G to be linear uniquely determines control points L_1 and

R_1 by Equation (9.4). If, say, L_1 has already been fixed by a two-split step (L_1 thus corresponds to a control point X_i above), there are not enough degrees of freedom to solve R_1, since it is over-determined by Equation (9.4). Instead, we let E, F, and G be quadratic, which gives the most general constraints of Equation (9.2). We let $e_0 = e_1 = e_2$, $f_0 = f_1 = f_2$, and $g_0 = g_1 = g_2$, so that Equation (9.2) reduces to:

(9.7a)
$$C_0 = e_0(M_1 - M_0) + f_0(L_0 - M_0) + g_0(R_0 - M_0) = 0,$$

(9.7b)
$$C_1 = e_0(M_1 - M_0) + 2e_0(M_2 - M_1) + f_0(L_0 - M_0) +$$
$$2f_0(L_1 - M_1) + g_0(R_0 - M_0) + 2g_0(R_1 - M_1) = 0,$$

(9.7c)
$$C_2 = e_0(M_1 - M_0) + 2e_0(M_2 - M_1) + e_0(M_3 - M_2) +$$
$$f_0(L_0 - M_0) + 2f_0(L_1 - M_1) + f_0(L_2 - M_2) +$$
$$g_0(R_0 - M_0) + 2g_0(R_1 - M_1) + g_0(R_2 - M_2) = 0,$$

(9.7d)
$$C_3 = 2e_0(M_2 - M_1) + e_0(M_3 - M_2) + 2f_0(L_1 - M_1) +$$
$$f_0(L_2 - M_2) + 2g_0(R_1 - M_1) + g_0(R_2 - M_2) = 0,$$

(9.7e)
$$C_4 = e_0(M_3 - M_2) + f_0(L_2 - M_2) + g_0(R_2 - M_2) = 0.$$

Equations (9.7a) and (9.7e) imply:

$$\frac{area(M_0, M_1, L_0)}{area(M_0, M_1, R_0)} = \frac{area(M_3, M_2, L_2)}{area(M_3, M_2, R_2)} = \frac{g_0}{f_0}.$$

An algorithm to set M_i, $i = 0, \ldots, 4$, L_0, L_2, R_0, and R_1 so as to satisfy this ratio is given by [Farin, 83]. Coefficients e_0, f_0, and g_0 are then determined up to a common factor, so that we can arbitrarily set $e_0 = 1$. Subtracting Equation (9.7a) from (9.7b) (or (9.7e) from (9.7d)) gives:

$$2e_0(M_2 - M_1) + 2f_0(L_1 - M_1) + 2g_0(R_1 - M_1) = 0,$$

from which we can determine R_1 because all other variables are known. Note that Equation (9.7c) is automatically satisfied: (9.7c)=(9.7a)+(9.7d)=(9.7b)+(9.7e).

So far, it has been essential that the two two-split triangles have four three-split triangles as neighbors. The triangles that are split into two can therefore not be chosen arbitrarily. Furthermore, the two-split triangles should be constructed before the three-split triangles, because L_1 in Step 6 (corresponding to a X_i in Step 5) must be known first.

To be able to adaptively choose which pairs of macro triangles are to be split into two, independently of the neighboring macro triangles, it should be possible to achieve tangent plane continuity when two or three sides of a triangle are split. This is indeed possible, and in fact gives even more degrees of freedom, but symmetry is lost and the determination of the control points gets a bit awkward.

9.8   A cubic six-split scheme

In the previous section we saw that the two-split scheme gives enough degrees of freedom to apply an optimization criterion to the solution of the tangent plane continuity constraints on the control points. We have also seen in Section 9.6 that such an optimization is important in order to avoid very distorted control point configurations, which result in tangent planes that are too tilted to be esthetically pleasing. However, the two-split as presented in Section 9.7 cannot be applied to all macro triangles.

A triangle split that provides sufficient degrees of freedom for optimization to be applied to all macro triangles is a split into six micro triangles. Figure 9.7 shows a schematic representation of a six-split of two adjacent cubic Bézier triangles, and the control point naming used in this section.

Let us consider the micro triangle edge between M_0 and M_3, and choose E, F, and G constant: $E = e$, etc. The tangent plane continuity condition, Equation (8.10), then is $t^2 C_0 + tu C_1 + u^2 C_2 = 0$, with C_0, C_1, and C_2 given by Equation (9.5), i.e.

$$(9.8a) \qquad C_0 = e(M_1 - M_0) + f(L_0 - M_0) + g(R_0 - M_0) = 0,$$

$$(9.8b) \qquad C_1 = e(M_2 - M_1) + f(L_1 - M_1) + g(R_1 - M_1) = 0,$$

$$(9.8c) \qquad C_2 = e(M_3 - M_2) + f(L_2 - M_2) + g(R_2 - M_2) = 0.$$

An alternative formulation but equivalent condition is the following:

$$(9.9a) \qquad \tilde{e}_1(M_0 - M_1) + \tilde{f}_1(L_0 - M_1) + \tilde{g}_1(R_0 - M_1) = 0,$$

$$(9.9b) \qquad \tilde{e}_1(M_1 - M_2) + \tilde{f}_1(L_1 - M_2) + \tilde{g}_1(R_1 - M_2) = 0,$$

$$(9.9c) \qquad \tilde{e}_1(M_2 - M_3) + \tilde{f}_1(L_2 - M_3) + \tilde{g}_1(R_2 - M_3) = 0.$$

Equations (9.9b) and (9.9c) in terms of X_i, Y_i, and Z, and applied to the four edges together, become

$$(9.10a) \qquad \tilde{e}_i(W_i - Y_i) + \tilde{f}_i(X_i - Y_i) + \tilde{g}_i(X_{i+1} - Y_i) = 0,$$

$$(9.10b) \qquad \tilde{e}_i(Y_i - Z) + \tilde{f}_i(Y_{i-1} - Z) + \tilde{g}_i(Y_{i+1} - Z) = 0,$$

with $i = 0, \ldots, 3$. For $i = 0$, Equation (9.10a) can be written as

$$(9.11a) \qquad \tilde{e}_i(Q_i - N_i) + \tilde{f}_i(P_{i-1} - N_i) + \tilde{g}_i(P_i - N_i) = 0.$$

For tangent plane continuity along the whole micro triangle edge between Z and O, the following condition must also hold:

$$(9.11b) \qquad \tilde{e}_i(O - Q_i) + \tilde{f}_i(Q_{i-1} - Q_i) + \tilde{g}_i(Q_{i+1} - Q_i) = 0.$$

Similar conditions apply to the edges corresponding to N_2, Q_2 and N_4, Q_4. For $i = 0, 2, 4$ the $\tilde{e}_i$ will be given the same value, say h, and likewise $\tilde{f}_i = k$, and $\tilde{g}_i = \ell$ for $i = 0, 2, 4$. Equation (9.11) thus becomes

$$(9.12a) \qquad h(Q_i - N_i) + k(P_i - N_i) + \ell(P_{i+1} - N_i) = 0,$$

$$(9.12b) \qquad h(O - Q_i) + k(Q_{i-1} - Q_i) + \ell(Q_{i+1} - Q_i) = 0.$$

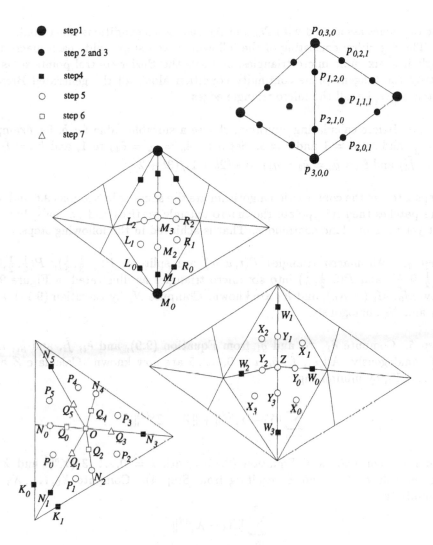

Figure 9.7. Control points used in the six-split scheme.

For $i = 1, 3, 5$ similar conditions must be satisfied, but the constants may be different, say a, b, and c:

$$(9.13a) \qquad a(Q_i - N_i) + b(P_i - N_i) + c(P_{i+1} - N_i) = 0,$$
$$(9.13b) \qquad a(O - Q_i) + b(Q_{i-1} - Q_i) + c(Q_{i+1} - Q_i) = 0.$$

With $i = 1$, Equation (9.13) applies to the edge between M_0 and O. In order to achieve tangent plane continuity along this whole micro triangle edge, the following condition must also hold:

$$(9.14) \qquad a(N_1 - M_0) + b(K_0 - M_0) + c(K_1 - M_0) = 0.$$

For the edges associated with N_3 and N_5 analogous conditions must hold.

The algorithm consisting of the following seven steps splits each macro triangle into six cubic micro triangles, and sets the Bézier control points so as to satisfy the tangent plane continuity condition along all the generated Brezier patch edges, i.e. all the micro triangle edges.

Step 0. Before calculating anything, choose a suitable value for b, for example $b = \frac{1}{2}$, and set $a = 1$ and $c = b$. Set h ($= \tilde{e}_0 = \tilde{e}_2 = \tilde{e}_4$) to 1, and k ($= \tilde{f}_0 = \tilde{f}_2 = \tilde{f}_4$) and ℓ ($= \tilde{g}_0 = \tilde{g}_2 = \tilde{g}_4$) to $-(2b+1)/(3b+2)$.

Steps 1 to 3 of the construction algorithm are the same as in Sections 9.6 and 9.7. The patches then interpolate the macro triangle vertices and normals, but are not yet tangent plane continuous. That is achieved in the following steps.

Step 4. All macro triangles $P(t, u, w)$ are split at $P(\frac{1}{3}, \frac{1}{3}, \frac{1}{3})$, $P(\frac{1}{2}, \frac{1}{2}, 0)$, $P(\frac{1}{2}, 0, \frac{1}{2})$, and $P(0, \frac{1}{2}, \frac{1}{2})$ into six micro triangles as illustrated in Figure 9.7. Now M_0, M_1 ($= K_0$), and K_1 are known. Compute N_1 by Equation (9.14), and N_3 and N_5 analogously.

Step 5. Compute $\tilde{e}_1$, $\tilde{f}_1$, and $\tilde{g}_1$ from Equation (9.9), and $\tilde{e}_i$, $\tilde{f}_i$, and $\tilde{g}_i$, $i = 3, 5$, analogously. All $\tilde{e}_i$, $\tilde{f}_i$, $\tilde{g}_i$, $i = 0, \dots, 5$ are now known. Calculate Z and $Y_0, \dots, Y_3$, by minimizing

$$\sum_{i=0}^{3} \|Y_i - Y_i^{old}\| + \|Z - Z^{old}\|,$$

under the constraints of Equation (9.10b) with $i = 0, \dots, 3$ (Y_i^{old} and Z^{old} are the values of Y_i and Z resulting from Step 4). Compute $X_0, \dots, X_3$ by minimizing

$$\sum_{i=0}^{3} \|X_i - X_i^{old}\|,$$

under the constraints of Equation (9.10a) with $i = 1, 3$ (X_i^{old} is the value of X_i resulting from Step 4).

Step 6. All P_i and N_i are known by now. Compute Q_0, Q_2, and Q_4 by Equation (9.12a) with $i = 0, 2, 4$. Set $O = (Q_0 + Q_2 + Q_4)/3$.

Step 7. Compute Q_1, Q_3, and Q_5 by Equation (9.13a) with $i = 1, 3, 5$.

It is not immediately clear that the above algorithm gives an overall tangent plane continuous surface. This is proved by the following theorem:

THEOREM 9.2 *Steps 0 to 7 above construct an overall tangent plane continuous surface.*

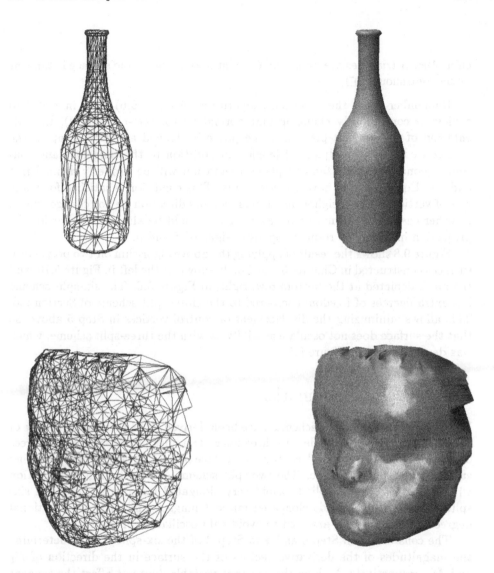

Figure 9.8. Left: input triangulations. Right: G^1-continuous surfaces resulting from the six-split scheme.

Proof. By construction (Step 5) the micro triangle edges between $p_{3,0,0}$ and Z, and between $p_{0,3,0}$ and Z satisfy Equation (9.8). The surface is thus tangent plane continuous along these edges.

The edges incident to O satisfy the tangent plane continuity condition Equation (8.10) if there is no conflict around O, that is, if both Equation (9.12b) for $i = 0, 2, 4$, and Equation (9.13b) for $i = 1, 3, 5$ are satisfied. The crux of the algorithm is that both conditions are automatically satisfied by setting O to $(Q_0 + Q_2 + Q_4)/3$ and by the special relation between b and k: $k = -(2b + 1)/(3b + 2)$. Incidentally, O is also equal to $(Q_1 + Q_3 + Q_5)/3$.

The surface is thus tangent plane continuous along all triangle edges. Since cubic Bézier triangles are internally C^2-continuous, the whole surface is tangent plane continuous. □

Remember from the three-split algorithm (Section 9.6) that in order to achieve a continuously changing unit normal vector (G^1-continuity), the orientation of the tangent plane must be properly defined at every point on the common edge of two adjacent triangles, in addition to the tangent plane continuity condition. The tangent plane orientation will be properly defined if f and g in Equation (9.8) have opposite signs. This condition is satisfied for many sets of vertices whose neighboring normal vectors direction do not change wildly. In other cases, the normal vector at a vertex should be altered in order for the projection in Step 2 to result in opposite signs of f and g.

Figure 9.8 shows the result of applying the above algorithm on two polyhedral surfaces constructed in Chapter 5: the bottle shown at the left in Figure 5.10, and the mask depicted at the bottom row, right, in Figure 5.6. The six-split scheme has extra degrees of freedom compared to the three-split scheme of Section 9.6. This allows minimizing the displacement of control vertices in Step 5 above, so that the surface does not oscillate as wildly as with the three-split scheme, which was demonstrated in Figure 9.5.

9.9   Concluding remarks

In this chapter, three-split schemes have been derived. The three-split scheme of subdividing a Bézier triangle into three micro triangles provides enough degrees of freedom to construct a G^1-continuous surface in a local way, but the resulting surface may wildly oscillate. The two-split scheme can be used in combination with the triangle three-split to avoid very elongated micro triangles. The six-split algorithm also avoids elongated micro triangles, and provides additional degrees of freedom that are used to avoid wild oscillation of the surface.

The constants b in Step 0 and $\frac{1}{3}$ in Step 3 of the six-split scheme determine the magnitudes of the derivative vectors of the surface in the direction of N_1 and M_1, respectively. Making the constant variable does not affect the tangent plane continuity of the surface. Rather, the constants become parameters that act like tension parameters at the vertices of the macro triangle.

The algorithms presented in Sections 9.6, 9.7, and 9.8 all have a time complexity $\mathcal{O}(N_e) = \mathcal{O}(N_t)$. This is clear from the iterations over the macro triangles and their edges, and the relation between the number of triangles and edges: $3N_t = 2N_e$.

A G^1-continuous surface generally looks smooth, but its appearance depends on the illumination. As has been derived in Chapter 8, the reflection of a linear light source on a G^1-continuous surface need not be G^1. For G^2-continuity more control points are needed than we have used here. This means that rational polynomial patches [Hagen and Pottmann, 88] or higher degree patches [Hogervorst and Damme, 92] must be used in order to keep the scheme local.

10

Conclusions

The main results of the research described in this thesis are summarized below.

Chapter 3. The γ-neighborhood graph is a new parameterized geometric graph. By its two parameters, a whole family of geometric graphs is defined, ranging from the empty to the complete graph. For particular choices of the parameters, the γ-graph reduces to known graphs such as the Convex Hull, the Gabriel Graph, the β_c-Skeleton, and the Delaunay Triangulation. The γ-graph unifies these graphs into a continuous spectrum.

Chapter 5. The geometric information contained in the γ-graph is used to construct a closed piecewise linear object boundary from scattered points. The γ-graph on the set of points is successively constricted until the boundary of the pruned γ-graph is a proper object boundary, passing through all the vertices. While constriction of the Delaunay Triangulation may stop unsuccessfully, the parameters of the γ-graph provide the flexibility to find a boundary through all the vertices. The use of the geometric information in the γ-graph by means of the γ-indicator results in good looking boundaries.

Chapter 7. The flintstone scheme is both an approximation and a localization scheme, and is hierarchical. This scheme can be applied to the constructed polygonal or polyhedral boundaries. Its definition is based on discs or balls, which makes the representation storage efficient, and hierarchical operations, for example intersections, computationally cheap.

Chapter 9. Given a polygonal boundary with (estimated) tangent vectors at the vertices, or a closed polyhedral surface with normal vectors at the vertices, a G^1-continuous piecewise cubic Bézier boundary is constructed in a local way. The six-split algorithm to subdivide a Bézier triangle into six micro triangles avoids thin triangles, and provides sufficient degrees of freedom to apply an optimization in order to prevent severe oscillations of the G^1-continuous surface.

Care has been taken to introduce new concepts that naturally generalize from 2D to 3D. This is exhibited by the definition of the γ-graph, the deletion rule in the constriction algorithm, and the definition of the flintstone scheme. (Chapter 9 introduced a new solution to the problem of interpolating vertices and surface normal vectors rather than a new concept.)

The major new algorithms introduced in this thesis have been implemented in C or C++, and their graphical results are shown in the successive chapters. The graphics rendering has been implemented on a variety of workstation platforms:

- IBM's RT with the 5080 graphics device, programmed in its assembly language [Veltkamp, 87].

- the RT with the Megapel graphics device, programmed in PHIGS;

- Sun's Sun3 with the TAAC-1 graphics device, programmed in C;

- Sun's SparcStation 1+ with the apE environment, software for visualization from the Ohio State University supercomputer center, 'programmed' visually;

- the Iris 4D VGX of Silicon Graphics, with their visualization software Explorer, also 'programmed' visually.

The pictures that show the 2D results of the algorithms and the 3D stereographic γ-neighborhood graph in Figure 3.7 are made with Adobe's Postscript. The rest of the pictures that show the 3D results are made by screen dumps. Because the implementation of the algorithms and the visualization took place on such different platforms, the various software modules are separate and have not been integrated.

Ideally, the algorithms should be interactive, so that the user can:

- adapt the parameters of the γ-graph, and dynamically add vertices to the graph,

- control the deletion of boundary tetrahedra when all vertices already lie on the boundary, and indicate boundary segments that may not be deleted in the constrained constriction procedure,

- adapt the level of approximation and the maximum approximation error in the flintstone scheme, or dynamically add vertices to the objects being approximated,

- alter the 'tension' parameters in Steps 0 and 2 of the six-split scheme, and the programs should show the result immediately. Except for the optimization step in the six-split scheme, all the algorithms themselves are fast enough to do this, even for large data sets. It is the data transfer from or to the computer's hard-disc or the graphics board that may be the bottleneck. Although the steering of the algorithms can be done interactively, the whole boundary construction process is far less interactive than the approach taken by [Veltkamp, 85]. There, 3D boundaries are constructed by interactively deforming primitive objects, such as tubes and sphere-like objects, so as to fit the given data. These deformable objects are represented by B-spline surfaces and are deformed by high level operators such as 'squeeze' and 'bend', which internally operate on the B-spline control points.

Possible future research directions are mentioned at the end of Chapters 3, 5, 7, and 9. Several topics treated in these chapters can be combined to yield other research directions:

- The boundary construction procedure presented in Chapter 5 constricts the Convex Hull until the boundary of the γ-graph passes through all vertices. All the intermediate boundaries can be considered approximations of the final boundary. All these approximations lie outside the final boundary. The opposite of a constriction process is a swell process, which starts with a polygon or polyhedron and expands these until all the vertices are included, in such a way that none of the vertices lie in the interior of the polygon or polyhedron. The intermediate boundaries are then approximations lying inside the final boundary. The inner and outer approximations together define a bounding area or volume of the boundary.

- Another research subject is the combination of approximation and the construction of a smooth boundary. We can construct curved boundary segments that interpolate part of the given vertices and approximate the rest of the vertices, and refine those segments whose approximation error is too large. A similar approach is taken by [Schmitt et al., 86] for *rectangular* patches on a *regular* grid, while our application has an arbitrary topology with triangular patches.

References

[Ahuja, 82] N. Ahuja. Dot pattern processing using Voronoi neighborhoods. *IEEE Transactions on Pattern Analysis and Machine Intelligence*, PAMI-4(3), 1982, 336 – 343.

[Alevizos et al., 87] P. D. Alevizos, J. D. Boissonnat, and M. Yvinec. An optimal $o(n \log n)$ algorithm for contour reconstruction from rays. In *Proceedings of the 3rd ACM Symposium on Computational Geometry*, ACM Press, 1987, 162 – 170.

[Ballard, 81] D. H. Ballard. Strip trees: a hierarchical representation for curves. *Communications of the ACM*, 24(5), 1981, 310 – 321.

[Barnhill, 85] R. E. Barnhill. Surfaces in computer aided geometric design: a survey with new results. *Computer Aided Geometric Design*, 2, 1985, 1 – 17.

[Barnhill and Böhm, 83] R. E. Barnhill and W. Böhm (editors). *Surfaces in Computer Aided Geometric Design*, North-Holland, 1983.

[Barsky and Beatty, 83] B. A. Barsky and J. C. Beatty. Local control of bias and tension in beta-splines. *ACM Transactions on Graphics*, 2(2), 1983, 109 – 134.

[Bentley and Shamos, 76] J. L. Bentley and M. I. Shamos. Divide-and-conquer in multidimensional space. In *Proceedings of the 8th Annual Symposium on Theory of Computing*, 1976, 220 – 230.

[Bernroider, 78] G. Bernroider. The foundation of computational geometry: theory and application of the point-lattice-concept within modern structure analysis. In R. E. Miles and J. Serra (editors), *Geometrical Probability and Biological Structures*, Springer-Verlag, 1978, 153 – 170.

[Bernstein, 12] S. Bernstein. Démonstration de théorème de Weierstrass fondeé sur le calcul des probabilités. *Harkov Soobs. Matem ob-va*, 13(1–2), 1912.

[Böhm, 85] W. Böhm. Curvature continuous curves and surfaces. *Computer Aided Geometric Design*, 2, 1985, 313 – 323.

[Böhm and Farin, 83] W. Böhm and G. Farin. Letter to the editor. *Computer Aided Design*, 15(5), 1983, 260 – 261.

[Boissonnat, 82] J.-D. Boissonnat. Representation of objects by triangulating points in 3-D space. In *Proceedings of the 6th International Conference on Pattern Recognition*, 1982, 830 – 832.

[Boissonnat, 84a] J.-D. Boissonnat. Geometric structures for three-dimensional shape representation. *ACM Transactions on Graphics*, 3(4), 1984, 266 – 286.

[Boissonnat, 84b] J.-D. Boissonnat. Representing 2D and 3D shapes with the Delaunay triangulation. In *Proceedings of the 7th International Conference on Pattern Recognition*, 1984, 745 – 748.

[Boissonnat, 88] J.-D. Boissonnat. Shape reconstruction from planar cross sections. *Computer Vision, Graphics, and Image Processing*, 44, 1988, 1 – 29.

[Boissonnat and Tellaud, 86] J.-D. Boissonnat and M. Tellaud. A hierarchical representation of objects: the Delaunay tree. In *Proceedings of the 2nd ACM Symposium on Computational Geometry*, ACM Press, 1986, 260 – 268.

[Bollobás, 79] B. Bollobás. *Graph Theory, An Introductory Course*. Springer-Verlag, 1979.

[Brown, 79] K. Q. Brown. Voronoi diagrams from convex hulls. *Information Processing Letters*, 9(5), 1979, 223 – 228.

[Casale, 87] M. S. Casale. Free-form surface modelling with trimmed surface patches. *IEEE Computer Graphics & Applications*, 7(1), 1987, 33 – 43.

[Chazelle, 91] B. Chazelle. An optimal convex hull algorithm and new results on cuttings. In *Proceedings of the 32nd Annual Symposium on Foundations of Computer Science*, IEEE Computer Society Press, 1991, 29 – 38.

[Choi et al., 88] B. K. Choi, H. Y. Shin, Y. I. Yoon, and J. W. Lee. Triangulation of scattered data in 3D space. *Computer Aided Design*, 20(5), 1988, 239 – 248.

[Clark, 76] J. H. Clark. Hierarchical geometric models for visible surface algorithms. *Communications of the ACM*, 19(10), 1976, 547 – 554.

[Clough and Tocher, 65] R. W. Clough and J. L. Tocher. Finite element stiffness matrices for analysis of plates in blending. In *Proceedings of the Conference on Matrix Methods in Structural Mechanics*, Air Force Institute of Technology, Wright-Patterson A. F. B., Ohio, 1965.

[Corby and Mundy, 90] N. R. Corby and J. L. Mundy. Applications of range image sensing and processing. In R. C. Jain and A. K. Jain (editors), *Analysis and Interpretation of Range Images*, Springer-Verlag, 1990, 255 – 272.

[Cottin and Damme, 90] C. Cottin and R. van Damme. 3D reconstruction of closed objects by piecewise cubic triangular Bézier patches. Technical Report 885, University of Twente, Enschede, The Netherlands, 1990. To be published in J. C. Mason and M. G. Cox (editors), *Mathematics of Surfaces III*.

[DeFloriani, 89] L. DeFloriani. A pyramidal data structure for triangle-based surface description. *IEEE Computer Graphics & Applications*, 9(2), 1989, 67 – 80.

[Delaunay, 28] B. Delaunay. Sur la sphère vide. In *Proceedings of the International Congress on Mathematics (Toronto 1924)*, Volume 1, University of Toronto Press, 1928, 695 – 700.

[Delaunay, 34] B. Delaunay. Sur la sphère vide. *Izvestija Akademii Nauk S.S.S.R. Otdelenie Matematiceskich i Estestvennych Nauk (Bulletin de l'Académie des Sciences de l'URSS, VII Série, Classe des Sciences Mathématiques et Naturelles)*, 1934, 793 – 800.

[DeRose and Barsky, 85] T. DeRose and B. A. Barsky. An intuitive approach to geometric continuity for parametric curves and surfaces. In N. Magnenat-Thalmann and D. Thalmann (editors), *Computer-Generated Images – The State of the Art*, Springer-Verlag, 1985, 159 – 175.

[DeRose, 90] T. D. DeRose. Necessary and sufficient conditions for tangent plane continuity of Bézier surfaces. *Computer Aided Geometric Design*, 7, 1990, 165 – 179.

[Devroye, 88] L. Devroye. The expected size of some graphs in computational geometry. *Computers & Mathematics with Applications*, 15(1), 1988, 53 –64.

[Dijkstra, 59] E. W. Dijkstra. A note on two problems in connection with graphs. *Numerische Mathematik*, 1, 1959, 269 – 271.

[Dillencourt, 87] M. B. Dillencourt. A non-Hamiltonian, nondegenerate Delaunay triangulation. *Information Processing Letters*, 25(3), 1987, 149 – 151.

[Dillencourt, 89] M. B. Dillencourt. An upper bound on the shortest exponent of inscribable polytopes. *Journal of Combinatorial Theory, Series B*, 46(1), 1989, 66 – 83.

[Dirac, 72] G. A. Dirac. On Hamiltonian circuits and Hamiltonian paths. *Mathematische Annalen*, 197, 1972, 57 – 70.

[Dominguez and Günther, 91] S. Dominguez and O. Günther. Performance analysis of three curve representation schemes. In H. Bieri and H. Noltemeier (editors), *Computational Geometry – Methods, Algorithms and Applications, Proceedings of the International Workshop on Computational Geometry CG'91, Bern, Switzerland*, Volume 553 of *Lecture Notes in Computer Science*, Springer-Verlag, 1991, 37 – 56.

[Douglas and Peucker, 73] D. H. Douglas and T. K. Peucker. Algorithms for the reduction of the number of points required to represent a digitized line or its caricature. *The Canadian Cartographer*, 10(2), 1973, 112 – 122.

[Duda and Hart, 73] R. O. Duda and P. E. Hart. *Pattern Classification and Scene Analysis*. John Wiley & Sons, 1973.

[Dwyer, 89] R. A. Dwyer. Higher-dimensional Voronoi diagrams in linear expected time. In *Proceedings of the ACM Symposium on Computational Geometry*, ACM Press, 1989, 326 – 333.

[Edelsbrunner, 87] H. Edelsbrunner. *Algorithms in Combinatorial Geometry*. Springer-Verlag, 1987.

[Edelsbrunner et al., 83] H. Edelsbrunner, D. G. Kirkpatrick, and R. Seidel. On the shape of a set of points in the plane. *IEEE Transactions on Information Theory*, IT-29(4), 1983, 551 – 559.

[Ekoule et al., 91] A. B. Ekoule, F. C. Peyrin, and L. Odet. A triangulation algorithm from arbitrary shaped multiple planar contours. *ACM Transactions on Graphics*, 10(2), 1991, 182 – 199.

[Farin, 82a] G. Farin. A construction for visual C^1 continuity of polynomial surface patches. *Computer Graphics and Image Processing*, 20(3), 1982, 272 – 282.

[Farin, 82b] G. Farin. Visually C^2 cubic splines. *Computer Aided Design*, 14(3), 1982, 137 – 139.

[Farin, 83] G. Farin. Smooth interpolation to scattered 3D data. In [Barnhill and Böhm, 83], 43 – 63.

[Farin, 86] G. Farin. Triangular Bernstein-Bézier patches. *Computer Aided Geometric Design*, 3(2), 1986, 83 – 127.

[Farin, 87] G. Farin (editor). *Geometric Modeling: Algorithms and New Trends*, SIAM, 1987.

[Farin, 90a] G. Farin. *Curves and Surfaces for Computer Aided Geometric Design*, 2nd edition. Academic Press, 1990.

[Farin, 90b] G. Farin. Surfaces over Dirichlet tessellations. *Computer Aided Geometric Design*, 7, 1990, 281 – 292.

[Faugeras et al., 84] O. D. Faugeras, M. Hebert, P. Mussi, and J. D. Boissonnat. Polyhedral approximation of 3-D objects without holes. *Computer Vision, Graphics, and Image Processing*, 25, 1984, 169 – 183.

[Foley et al., 90] J. D. Foley, A. van Dam, S. K. Feiner, and J. F. Hughes. *Computer Graphics: Principles and Practice*, 2nd edition. Addison-Wesley, 1990.

[Forrest, 71] A. R. Forrest. Computational geometry. In *Proceedings of the Royal Society London A.*, 1971, 187 – 195.

[Fuchs et al., 77] H. Fuchs, Z. M. Kedem, and S. P. Uselton. Optimal surface reconstruction from planar contours. *Communications of the ACM*, 20(10), 1977, 683 – 702.

[Gabriel and Sokal, 69] K. R. Gabriel and R. R. Sokal. A new statistical approach to geographic variation analysis. *Systematic Zoology*, 18, 1969, 259 – 278.

[Geise, 62] G. Geise. Über berührende Kegelschnitte einer ebenen Kurve. *Zeitschrift für Angewandte Mathematik und Mechanik*, 42(7/8), 1962, 297 – 304.

[Goldman, 83] R. N. Goldman. Subdivision algorithms for Bézier triangles. *Computer Aided Design*, 15(3), 1983, 159 – 166. See also [Böhm and Farin, 83].

[Günther, 88] O. Günther. *Efficient Structures for Geometric Data Management*, Volume 337 of *Lecture Notes in Computer Sciences*. Springer-Verlag, 1988.

[Hagen and Pottmann, 88] H. Hagen and H. Pottmann. Curvature continuous triangular interpolants. In T. Lyche and L. L. Schumaker (editors), *Mathematical Methods in Computer Aided Geometric Design (conference held in Olso, Norway, 1988)*, Academic Press, 1988, 373 – 384.

[Haralick and Shapiro, 92] R. M. Haralick and L. G. Shapiro. *Computer and Robot Vision, Volume 1*. Addison-Wesley, 1992.

[Herron, 85] G. Herron. Smooth closed surfaces with discrete triangular interpolants. *Computer Aided Geometric Design*, 2(4), 1985, 297 – 306.

[Hershberger and Snoeyink, 92] J. Hershberger and J. Snoeyink. Speeding up the Douglas-Peucker line simplification algorithm. In *Proceedings of the 5th International Symposium on Spatial Data Handling*, IGU Commission on GIS, Charleston, South Carolina, 1992, 134 – 143.

[Higashi et al., 90] M. Higashi, T. Kushimoto, and M. Hosaka. On formulation and display for visualizing features and evaluating quality of free-form surfaces. In C. E. Vandoni and D. A. Duce (editors), *Eurographics '90*, North-Holland, 1990, 299 – 309.

[Hogervorst and Damme, 92] B. J. Hogervorst and R. van Damme. Degenerate polynomial patches of degree 11 for almost GC^2 interpolation over triangles. In *Proceedings of the 3rd International Conference on Algorithms for Approximation, Oxford, U.K.*, 1992.

[Imai and Iri, 88] H. Imai and M. Iri. Polygonal approximations of a curve – formulations and algorithms. In [Toussaint, 88a], 71 – 86.

[Jensen, 87] T. Jensen. Assembling triangular and rectangular patches and multivariate splines. In [Farin, 87], 203 – 220.

[Keppel, 75] E. Keppel. Approximating complex surfaces by triangulation of contour lines. *IBM Journal of Research and Development*, 19(1), 1975, 2 – 11.

[Kirkpatrick and Radke, 85] D. G. Kirkpatrick and J. D. Radke. A framework for computational morphology. In G. T. Toussaint (editor), *Computational Geometry*, Elsevier Science Publishers, 1985, 217 – 248.

[Klee, 80] V. Klee. On the complexity of d-dimensional Voronoi diagrams. *Archiv der Mathematik*, 34, 1980, 75 – 80.

[Knuth, 76] D. E. Knuth. Big omicron and big omega and big theta. *SIGACT News*, 8(2), 1976, 18 – 24.

[Koenderink, 90] J. J. Koenderink. *Solid Shape*. MIT Press, 1990.

[Lankford, 69] P. M. Lankford. Regionalization: theory and alternative algorithms. *Geographical Analysis*, 1(2), 1969, 169 – 212.

[Lawson, 77] C. L. Lawson. Software for C^1 surface interpolation. In J. R. Rice (editor), *Mathematical Software III*, Academic Press, 1977, 161 – 194.

[Lee and Schachter, 80] D. T. Lee and B. J. Schachter. Two algorithms for constructing the Delaunay triangulation. *International Journal of Computers and Information Science*, 9(3), 1980, 219 – 242.

[Liu and Hoschek, 89] D. Liu and J. Hoschek. CG^1 continuity conditions between adjacent rectangular and triangular Bézier surface patches. *Computer Aided Design*, 21(4), 1989, 194–200.

[Manning, 74] J. R. Manning. Continuity conditions for spline curves. *The Computer Journal*, 17(2), 1974, 181 – 186.

[Mäntylä, 88] M. Mäntylä. *An Introduction to Solid Modeling*. Computer Science Press, 1988.

[Matula and Sokal, 80] D. W. Matula and R. R. Sokal. Properties of Gabriel graphs relevant to geographic variation research and the clustering of points in the plane. *Geographical Analysis*, 12, 1980, 205 – 222.

[Meagher, 82] D. Meagher. Geometric modeling using octree encoding. *Computer Graphics and Image Processing*, 19, 1982, 129 – 147.

[Medek, 81] V. Medek. On the boundary of a finite set of points in the plane.

Computer Vision, Graphics, and Image Processing, 15, 1981, 93 – 99.

[Megiddo, 83] N. Megiddo. Linear time algorithms for linear programming in $\mathbb{R}^3$ and related problems. *SIAM Journal on Computing*, 12(4), 1983, 759 – 776.

[Mehlhorn, 84] K. Mehlhorn. *Data Structures and Algorithms 2: Graph Algorithms and NP-Completeness*. Springer-Verlag, 1984.

[Miles, 70] R. E. Miles. On the homogeneous planar Poisson point process. *Mathematical Biosciences*, 6, 1970, 85 – 127.

[Minsky and Papert, 69] M. Minsky and S. Papert. *Perceptrons: An Introduction to Computational Geometry*. MIT Press, 1969.

[Nielson, 87] G. M. Nielson. A transfinite, visually continuous, triangular interpolant. In [Farin, 87], 235 – 246.

[Nielson and Franke, 83] G. M. Nielson and R. Franke. Surface construction based upon triangulations. In [Barnhill and Böhm, 83], 163 – 177.

[Oosterom and Bos, 89] P. van Oosterom and J. van den Bos. An object-oriented approach to the design of geographic information systems. *Computers & Graphics*, 13(4), 1989, 409 – 418.

[O'Rourke, 81] J. O'Rourke. Polyhedra of minimal area as 3D object models. In *Proceedings of the International Joint Conference on Artificial Intelligence*, 1981, 664 – 666.

[O'Rourke, 86] J. O'Rourke. The computational geometry column. *Computer Graphics*, 20(5), 1986, 232 – 234.

[O'Rourke and Badler, 79] J. O'Rourke and N. Badler. Decomposition of three-dimensional objects into spheres. *IEEE Pattern Analysis and Machine Intelligence*, PAMI-1(3), 1979, 295 – 305.

[O'Rourke et al., 87] J. O'Rourke, H. Booth, and R. Washington. Connect-the-dots: a new heuristic. *Computer Vision, Graphics, and Image Processing*, 39, 1987, 258 – 266.

[Pfluger and Neamtu, 91] P. Pfluger and M. Neamtu. Geometrically smooth interpolation by triangular Bernstein–Bézier patches with coalescent control points. In P. J. Laurent, A. L. Méhauté, and L. L. Schumaker (editors), *Curves and Surfaces*, Academic Press, 1991, 363 – 366.

[Phong, 75] B. T. Phong. Illumination for computer-generated pictures. *Communications of the ACM*, 18(6), 1975, 311 – 317.

[Piper, 87] B. R. Piper. Visually smooth interpolation with triangular Bézier patches. In [Farin, 87], 221 – 233.

[Ponce and Faugeras, 87] J. Ponce and O. Faugeras. An object centered hierarchical representation for 3D objects: the prism tree. *Computer Vision, Graphics, and Image Processing*, 38(1), 1987, 1 – 28.

[Preparata and Hong, 77] F. P. Preparata and S. J. Hong. Convex hulls of finite sets of points in two and three dimensions. *Communications of the ACM*, 20(2), 1977, 87 – 93.

[Preparata and Shamos, 85] F. P. Preparata and M. I. Shamos. *Computational Geometry, an Introduction*. Springer-Verlag, 1985.

[Prim, 57] R. C. Prim. Shortest connection networks and some generalizations. *Bell Systems Technical Journal*, 36, 1957, 1389 – 1401.

[QuaVis, 90] *AAAI-90 workshop on qualitative vision*. 1990.

[Ray and Ray, 92] B. K. Ray and K. S. Ray. An algorithm for polygonal approximation of digitized curves. *Pattern Recognotion Letters*, 13(7), 1992, 489 – 496.

[Rioux and Cournoyer, 88] M. Rioux and L. Cournoyer. The NRCC three-dimensional image data files. Technical Report CNRC 29077, National Research Council Canada, 1988.

[Samet, 84] H. Samet. The quadtree and related hierarchical data structures. *ACM Computing Surveys*, 16(2), 1984, 187 – 260.

[Schmitt and Gholizadeh, 86] F. Schmitt and B. Gholizadeh. Adaptive polyhedral approximation of digitized surfaces. *Proceedings of the SPIE – The International Society for Optical Engineering*, 595, 1986, (Proceedings of the SPIE Conference Computer Vision for Robots, Cannes, France, December 1985), 101 – 108.

[Schmitt et al., 86] F. J. M. Schmitt, B. Barsky, and W.-H. Du. An adaptive subdivision method for surfaces from sampled data. *Proceedings SIGGRAPH '86, Computer Graphics*, 20(4), 1986, 179 – 188.

[Schmitt et al., 91] F. J. M. Schmitt, X. Chen, and W.-H. Du. Geometric modeling from range image data. In F. H. Post and W. Barth (editors), *EURO-GRAPHICS'91*, 1991, 317 – 328.

[Schoenberg, 46] I. J. Schoenberg. Contributions to the problem of approximation of equidistant data by analytic functions. *Quarterly of Applied Mathematics*, 4, 1946, Part A: 45 – 99, Part B: 112 – 141.

[Seidel, 86] R. Seidel. Constructing higher-dimensional convex hulls at logarithmic cost per face. In *Proceedings of the 18th Annual ACM Symposium on Theory of Computing*, ACM Press, 1986, 404 – 413.

[Serra, 86] J. Serra. Introduction to mathematical morphology. *Computer Vision, Graphics, and Image Processing*, 35, 1986, 283 – 305.

[Shamos, 78] M. I. Shamos. *Computational Geometry*. PhD thesis, Department of Computer Science, Yale University, New Haven, Connecticut, 1978.

[Shamos and Hoey, 75] M. I. Shamos and D. Hoey. Closest point problems. In *Proceedings of the 16th Annual IEEE Symposium on Foundations of Computer Science*, IEEE, 1975, 151 – 162.

[Shirai, 87] Y. Shirai. *Three-Dimensional Computer Vision*. Springer-Verlag, 1987.

[Sloan, 91] K. Sloan. Surface normal (summary). *Usenet comp.graphics article*, 1991.

[Strang and Fix, 73] G. Strang and G. Fix. *An Analysis of the Finite Element Method*. Prentice-Hall, 1973.

[Su and Liu, 89] B.-Q. Su and D.-Y. Liu. *Computational Geometry — Curve and Surface Modeling*. Academic Press, 1989.

[Su and Chang, 91] T. Su and R. Chang. Computing the k-relative neighborhood graph in the Euclidean plane. *Pattern Recognition*, 24, 1991, 231 – 239.

[Supowit, 83] K. J. Supowit. The relative neighbourhood graph with an application to minimum spanning trees. *Journal of the ACM*, 30(3), 1983, 428 - 447.

[Thoenes, 84] C. Thoenes. Uccello's chalice. *Computer Aided Geometric Design*,

1, 1984, 97 – 99.

[Toussaint, 80] G. T. Toussaint. The relative neighbourhood graph of a finite planar set. *Pattern Recognition*, 12(4), 1980, 261 – 268.

[Toussaint, 88a] G. T. Toussaint (editor). *Computational Morphology – A Computational Geometric Approach to the Analysis of Form*. North-Holland, 1988.

[Toussaint, 88b] G. T. Toussaint. A graph theoretical primal sketch. In [Toussaint, 88a], 229 – 260.

[Tutte, 77] W. T. Tutte. Bridges and Hamiltonian circuits in planar graphs. *Aequationes Mathematicae*, 15, 1977, 1 – 33.

[Veltkamp, 85] R. C. Veltkamp. An interactive solid modeling approach to three-dimensional reconstruction. Master's thesis, Leiden University, Leiden, The Netherlands/IBM Scientific Center, Paris, France, 1985.

[Veltkamp, 87] R. C. Veltkamp. *RT/PC – 5080 Interface Reference Manual*. Leiden University, Leiden, The Netherlands, 1987.

[Veltkamp, 88] R. C. Veltkamp. The γ-neighbourhood graph for computational morphology. In *Proceedings of Computing Science in the Netherlands '88, Utrecht, The Netherlands*, 1988, 451 – 462.

[Veltkamp, 89a] R. C. Veltkamp. 2D and 3D computational morphology on the γ-neighborhood graph. *Acta Stereologica*, 8(2/2), 1989, 595 – 600.

[Veltkamp, 89b] R. C. Veltkamp. A divide-and-conquer algorithm to compute the 3D Delaunay triangulation. In P. M. G. Apers, D. Bosman, and J. van Leeuwen (editors), *Computing Science in the Netherlands '89, Utrecht, The Netherlands*, 1989, 463 – 480.

[Veltkamp, 90] R. C. Veltkamp. The flintstone representation and approximation scheme. In A. J. van de Goor (editor), *Computing Science in the Netherlands '90, Utrecht, The Netherlands*, 1990, 485 – 498.

[Veltkamp, 91] R. C. Veltkamp. 2D and 3D polygonal boundary reconstruction with the γ-neighborhood graph. Technical Report CS-R9116, CWI, Amsterdam, The Netherlands, 1991.

[Veltkamp, 92a] R. C. Veltkamp. Closed G^1-continuous cubic Bézier surfaces. Technical Report CS-R9226, CWI, Amsterdam, The Netherlands, 1992.

[Veltkamp, 92b] R. C. Veltkamp. The flintstones: hierarchical approximation and localization (extended abstract). In *Abstracts of the 8th European Workshop on Computational Geometry (CG'92), 12/13 March 1992, Utrecht, The Netherlands, Technical Report RUU-CS-92-10, Utrecht University, Utrecht, The Netherlands*, 1992, 69 – 72.

[Veltkamp, 92c] R. C. Veltkamp. The γ-neighborhood graph. *Computational Geometry, Theory and Applications*, 1(4), 1992, 227 – 246.

[Veltkamp, 92d] R. C. Veltkamp. Survey of continuities of curves and surfaces. *Computer Graphics Forum*, 11(2), 1992, 93 – 112.

[Voronoï, 08] G. Voronoï. Nouvelles applications des paramètres continus à la théorie des formes quadratiques. Deuxième mémoire — Recherche sur les parralléloèdres primitifs, Introduction et première partie. *Journal für die reine und angewandte Mathematik*, 134, 1908, 198 – 287.

Index

Springer-Verlag
and the Environment

We at Springer-Verlag firmly believe that an international science publisher has a special obligation to the environment, and our corporate policies consistently reflect this conviction.

We also expect our business partners – paper mills, printers, packaging manufacturers, etc. – to commit themselves to using environmentally friendly materials and production processes.

The paper in this book is made from low- or no-chlorine pulp and is acid free, in conformance with international standards for paper permanency.

Lecture Notes in Computer Science

For information about Vols. 1–808
please contact your bookseller or Springer-Verlag

Vol. 846: D. Shepherd, G. Blair, G. Coulson, N. Davies, F. Garcia (Eds.), Network and Operating System Support for Digital Audio and Video. Proceedings, 1993. VIII, 269 pages. 1994.

Vol. 847: A. L. Ralescu (Ed.) Fuzzy Logic in Artificial Intelligence. Proceedings, 1993. VII, 128 pages. 1994. (Subseries LNAI).

Vol. 848: A. R. Krommer, C. W. Ueberhuber, Numerical Integration on Advanced Computer Systems. XIII, 341 pages. 1994.

Vol. 849: R. W. Hartenstein, M. Z. Servít (Eds.), Field-Programmable Logic. Proceedings, 1994. XI, 434 pages. 1994.

Vol. 850: G. Levi, M. Rodríguez-Artalejo (Eds.), Algebraic and Logic Programming. Proceedings, 1994. VIII, 304 pages. 1994.

Vol. 851: H.-J. Kugler, A. Mullery, N. Niebert (Eds.), Towards a Pan-European Telecommunication Service Infrastructure. Proceedings, 1994. XIII, 582 pages. 1994.

Vol. 852: K. Echtle, D. Hammer, D. Powell (Eds.), Dependable Computing – EDCC-1. Proceedings, 1994. XVII, 618 pages. 1994.

Vol. 853: K. Bolding, L. Snyder (Eds.), Parallel Computer Routing and Communication. Proceedings, 1994. IX, 317 pages. 1994.

Vol. 854: B. Buchberger, J. Volkert (Eds.), Parallel Processing: CONPAR 94 – VAPP VI. Proceedings, 1994. XVI, 893 pages. 1994.

Vol. 855: J. van Leeuwen (Ed.), Algorithms – ESA '94. Proceedings, 1994. X, 510 pages.1994.

Vol. 856: D. Karagiannis (Ed.), Database and Expert Systems Applications. Proceedings, 1994. XVII, 807 pages. 1994.

Vol. 857: G. Tel, P. Vitányi (Eds.), Distributed Algorithms. Proceedings, 1994. X, 370 pages. 1994.

Vol. 858: E. Bertino, S. Urban (Eds.), Object-Oriented Methodologies and Systems. Proceedings, 1994. X, 386 pages. 1994.

Vol. 859: T. F. Melham, J. Camilleri (Eds.), Higher Order Logic Theorem Proving and Its Applications. Proceedings, 1994. IX, 470 pages. 1994.

Vol. 860: W. L. Zagler, G. Busby, R. R. Wagner (Eds.), Computers for Handicapped Persons. Proceedings, 1994. XX, 625 pages. 1994.

Vol: 861: B. Nebel, L. Dreschler-Fischer (Eds.), KI-94: Advances in Artificial Intelligence. Proceedings, 1994. IX, 401 pages. 1994. (Subseries LNAI).

Vol. 862: R. C. Carrasco, J. Oncina (Eds.), Grammatical Inference and Applications. Proceedings, 1994. VIII, 290 pages. 1994. (Subseries LNAI).

Vol. 863: H. Langmaack, W.-P. de Roever, J. Vytopil (Eds.), Formal Techniques in Real-Time and Fault-Tolerant Systems. Proceedings, 1994. XIV, 787 pages. 1994.

Vol. 864: B. Le Charlier (Ed.), Static Analysis. Proceedings, 1994. XII, 465 pages. 1994.

Vol. 865: T. C. Fogarty (Ed.), Evolutionary Computing. Proceedings, 1994. XII, 332 pages. 1994.

Vol. 866: Y. Davidor, H.-P. Schwefel, R. Männer (Eds.), Parallel Problem Solving from Nature - PPSN III. Proceedings, 1994. XV, 642 pages. 1994.

Vol 867: L. Steels, G. Schreiber, W. Van de Velde (Eds.), A Future for Knowledge Acquisition. Proceedings, 1994. XII, 414 pages. 1994. (Subseries LNAI).

Vol. 868: R. Steinmetz (Ed.), Multimedia: Advanced Teleservices and High-Speed Communication Architectures. Proceedings, 1994. IX, 451 pages. 1994.

Vol. 869: Z. W. Raś, Zemankova (Eds.), Methodologies for Intelligent Systems. Proceedings, 1994. X, 613 pages. 1994. (Subseries LNAI).

Vol. 870: J. S. Greenfield, Distributed Programming Paradigms with Cryptography Applications. XI, 182 pages. 1994.

Vol. 871: J. P. Lee, G. G. Grinstein (Eds.), Database Issues for Data Visualization. Proceedings, 1993. XIV, 229 pages. 1994.

Vol. 872: S Arikawa, K. P. Jantke (Eds.), Algorithmic Learning Theory. Proceedings, 1994. XIV, 575 pages. 1994.

Vol. 873: M. Naftalin, T. Denvir, M. Bertran (Eds.), FME '94: Industrial Benefit of Formal Methods. Proceedings, 1994. XI, 723 pages. 1994.

Vol. 874: A. Borning (Ed.), Principles and Practice of Constraint Programming. Proceedings, 1994. IX, 361 pages. 1994.

Vol. 875: D. Gollmann (Ed.), Computer Security – ESORICS 94. Proceedings, 1994. XI, 469 pages. 1994.

Vol. 876: B. Blumenthal, J. Gornostaev, C. Unger (Eds.), Human-Computer Interaction. Proceedings, 1994. IX, 239 pages. 1994.

Vol. 877: L. M. Adleman, M.-D. Huang (Eds.), Algorithmic Number Theory. Proceedings, 1994. IX, 323 pages. 1994.

Vol. 878: T. Ishida; Parallel, Distributed and Multiagent Production Systems. XVII, 166 pages. 1994. (Subseries LNAI).

Vol. 879: J. Dongarra, J. Waśniewski (Eds.), Parallel Scientific Computing. Proceedings, 1994. XI, 566 pages. 1994.

Vol. 880: P. S. Thiagarajan (Ed.), Foundations of Software Technology and Theoretical Computer Science. Proceedings, 1994. XI, 451 pages. 1994.

Vol. 881: P. Loucopoulos (Ed.), Entity-Relationship Approach – ER'94. Proceedings, 1994. XIII, 579 pages. 1994.

Vol. 882: D. Hutchison, A. Danthine, H. Leopold, G. Coulson (Eds.), Multimedia Transport and Teleservices. Proceedings, 1994. XI, 380 pages. 1994.

Vol. 883: L. Fribourg, F. Turini (Eds.), Logic Program Synthesis and Transformation – Meta-Programming in Logic. Proceedings, 1994. IX, 451 pages. 1994.

Vol. 884: J. Nievergelt, T. Roos, H.-J. Schek, P. Widmayer (Eds.), IGIS '94: Geographic Information Systems. Proceedings, 1994. VIII, 292 pages. 19944.

Vol. 885: R. C. Veltkamp, Closed Objects Boundaries from Scattered Points. VIII, 144 pages. 1994.

Vol. 886: M. M. Veloso, Planning and Learning by Analogical Reasoning. XIII, 181 pages. 1994. (Subseries LNAI).

Vol. 887: M. Toussaint (Ed.), Ada in Europe. Proceedings, 1994. XII, 521 pages. 1994.